Flora of Suffolk

A Catalogue of the Plants
(Indigenous or Naturalized)
Found in a Wild State
in the County of Suffolk

JOHN STEVENS HENSLOW
EDMUND SKEPPER

CAMBRIDGE
UNIVERSITY PRESS

CAMBRIDGE UNIVERSITY PRESS

Cambridge, New York, Melbourne, Madrid, Cape Town,
Singapore, São Paolo, Delhi, Mexico City

Published in the United States of America by Cambridge University Press, New York

www.cambridge.org
Information on this title: www.cambridge.org/9781108055673

© in this compilation Cambridge University Press 2013

This edition first published 1860
This digitally printed version 2013

ISBN 978-1-108-05567-3 Paperback

CAMBRIDGE LIBRARY COLLECTION

Books of enduring scholarly value

Life Sciences

Until the nineteenth century, the various subjects now known as the life sciences were regarded either as arcane studies which had little impact on ordinary daily life, or as a genteel hobby for the leisured classes. The increasing academic rigour and systematisation brought to the study of botany, zoology and other disciplines, and their adoption in university curricula, are reflected in the books reissued in this series.

Flora of Suffolk

An influential professor of botany at Cambridge, John Stevens Henslow (1796–1861) revived his department and helped develop the current University Botanical Garden for study, teaching and conservation. A mentor to the young Darwin, he proved an educational innovator, initiating the study of individual sciences at Cambridge and practical examinations at the University of London. While rector of Hitcham in Suffolk, he took an interest in local politics, welfare and popular education. This led to the publication in 1860 of this catalogue, which collated the observations and work of amateur botanists. Henslow was the overarching academic and technical consultant while Edmund Skepper is credited with organising and collating the information from the contributors. Catalogued taxonomically, each plant's Latin and common name is given along with its physical description, common locations, rarity or commonality, and periods of flowering or germination. It remains a valuable guide for amateur botanists and naturalists.

Cambridge University Press has long been a pioneer in the reissuing of out-of-print titles from its own backlist, producing digital reprints of books that are still sought after by scholars and students but could not be reprinted economically using traditional technology. The Cambridge Library Collection extends this activity to a wider range of books which are still of importance to researchers and professionals, either for the source material they contain, or as landmarks in the history of their academic discipline.

Drawing from the world-renowned collections in the Cambridge University Library and other partner libraries, and guided by the advice of experts in each subject area, Cambridge University Press is using state-of-the-art scanning machines in its own Printing House to capture the content of each book selected for inclusion. The files are processed to give a consistently clear, crisp image, and the books finished to the high quality standard for which the Press is recognised around the world. The latest print-on-demand technology ensures that the books will remain available indefinitely, and that orders for single or multiple copies can quickly be supplied.

The Cambridge Library Collection brings back to life books of enduring scholarly value (including out-of-copyright works originally issued by other publishers) across a wide range of disciplines in the humanities and social sciences and in science and technology.

FLORA OF SUFFOLK:

A

CATALOGUE OF THE PLANTS

(INDIGENOUS OR NATURALIZED,)

FOUND IN A WILD STATE

IN

THE COUNTY OF SUFFOLK;

BY THE

REV. J. S. HENSLOW, M.A., PROF. BOT.,

RECTOR OF HITCHAM;

AND

EDMUND SKEPPER,

BURY ST. EDMUND'S.

———————

" Consider the Lilies of the Field." Matt. vi. 28.

———————

LONDON:
SIMPKIN & MARSHALL, STATIONERS' HALL COURT.
BURY ST. EDMUND'S:
JACKSON AND FROST, ANGEL HILL.

PREFACE.

This Catalogue has been compiled from lists and notes furnished by the following persons, from various parts of the County of Suffolk: Mr. R. Blackett, (Bury): the Rev. E. N. Bloomfield, (Great Glemham and Sweffling, etc., etc.): Sir Charles F. Bunbury, Bart., (neighbourhood of Mildenhall): Mrs. Carss, Little Welnetham, etc.): Mr. Joseph Gedge, (neighbourhood of Bury): the Rev. J. S. Henslow, (Hitcham): the Rev. S. Rickards, (neighbourhood of Stowlangtoft): the Rev. Henry Roberts, (Naughton, etc.): Mr. E. Skepper, (Bury, Tuddenham, Lowestoft, etc.): Mr. D. Stock, (neighbourhood of Bungay): the Rev. K. Trimmer, (neighbourhood of Yarmouth): and Dr. White, (neighbourhood of Lavenham): in addition to which are localities from many districts by the late F. K. Eagle, Esq., and many more copied from "The Botanist's Guides," supplied by gentlemen whose services have already been acknowledged in the same Treatises.

Mr. Skepper has the entire merit of having reduced these materials to order, and of having seen them through the press. My part in the work has been that of a consulting, but, otherwise, sleeping partner. We had thought of saying something in regard to the geographic distribution of the species, but found our materials insufficient for treating this question to advantage. We have considered it best to put forth the "Catalogue" in its present crude state, in the hope of inviting many fellow-labourers to compile lists of the plants in their several localities. By interleaving a copy of the "Catalogue," additions can readily be inserted, and statements made respecting the geological formation, and superficial soil upon which each species grows. These memoranda will include the Chalk, Tertiary Sands and Clays, Coralline and Red Crags, Post-tertiary Clay and Gravel Drift. The maritime, marshy, boggy, heathy, and cultivated characters of the superficial soil should be added, because these frequently and completely mask the subjacent strata.

It is important, also, to state what common species
are rare or wanting in particular localities, and what
species, usually rare, may be common in those places.

Collections of dried specimens, or at least of all the
rarer species, should be prepared from each locality,
in order to secure absolute certainty that no error of
judgment has occurred in naming them. From these
specimens a selection might be made, to secure a
perfect set of Suffolk Plants for the Athenæum at
Bury, and another for the Museum at Ipswich.

With the "Handbook of the British Flora," by
Bentham, whose arrangement we have adopted in
the present list, beginners may soon learn to make
out the names of most of our wild flowers; the
" Key " introduced for this purpose being far pre-
ferable to the more usual plan, by the Linnean
System.

Having had some years' experience of the advantage
of introducing Botany as an *educational* weapon in a
humble village school, I can strongly recommend it
to all who are interested in raising the intellectual
status of our village children. Whoever may be desirous of seeing the plan we have adopted at Hitcham,
will find the children at their botanical exercises every

Monday at three o'clock. I have gradually accumu-
lated sundry memoranda, which I hope to find leisure,
before long, to throw into shape, that others may
be saved the mistakes I have made, and profit by the
experience I have acquired. It was for the sake of
facilitating our proceedings that I introduced the plan
of giving the Orders a common termination in "anths."
This plan has received the sanction of some eminent
Botanists, though it is distasteful to classical ears. I
greatly prefer it, where an English nomenclature is
adopted for the Genera and Species, to Mr. Bentham's plan of adding the word " Family " to the
typical genus. It is not more burdensome to the
memory to say " Ranunculanths," than " The Ra-
nunculus Family," and it is more in keeping with the
usual plan of systematists, who employ a single word
to express an Order. The Orders, thus named, have
been added to the Latin names in our " Catalogue,"
to assist any one who may feel desirous of following
the plan which has been adopted in the village school
at Hitcham. We have a printed list of our parish
wild flowers, in which the names of the Genera and
Species correspond to those employed by Bentham ;
and a copy of this is supplied to every child who is

desirous of joining the botanical class. I consider a twelvemonths' search among the wild flowers of any neighbourhood, ought to be sufficient to qualify a would-be teacher for starting a botanical class in a village school. Teacher and scholars may then go on improving together. Above all precautions, I advise a teacher not to *exact* an exercise more than one day in the week, but to allow any children, who may be desirous of getting on, to volunteer as much as they please on other days. A little tract, which I wrote two years ago for the Kensington Museum of Science and Art, explains the sort of exercises to which I allude.

With regard to the list of Acotyledons, this must be considered the most imperfect part of the "Catalogue," few localities having been diligently searched for these plants. Those which are recorded are generally upon the authority of Mr. Skepper, who has received assistance from the Rev. M. J. Berkeley, our great cryptogamic authority; some are contributions from the Rev. E. N. Bloomfield, Mr. D. Stock, and the late F. K. Eagle, Esq. They have been arranged according to the 5th Vol. of the "English Flora," edited by Sir W. J. Hooker and Dr. Arnott. The list viii.

of Diatomaceæ, (beginning with the genus Epithemia, at p. 116,) has received the valuable assistance of Mr. H. Taylor, of Ixworth ; and of the late Dr. Bleakeley, of Norwich, well known there for many years as a very able microscopist and botanist. It is arranged according to " Smith's Synopsis."

J. S. HENSLOW.

HITCHAM, *4th June,* 1860.

ABBREVIATIONS

*Used in place of the full names of Contributors, Treatises,
etc., quoted in the present Flora.*

J. T. List of Plants growing near Mildenhall, by
Mr. J. Townsend. Phytologist, Vol. 2.
Old Series.

W.L.N. List of Plants by Mr. W. L. Notcutt.
Phyt. Vol. 1. Old Series.

J.W.G. (misprinted J. W. G.) List of Plants by Mr.
T. W. Gissing, Wakefield. Phyt. New
Series.

A.N.H. Annals of Natural History. Fungi, by the
Rev. M. J. Berkeley.

O.B.G. Old Botanist's Guide. Messrs. Turner and
Dillwyn.

N.B.G. New Ditto. Mr. H. C. Watson.

Br. Fl. British Flora. Sir W. J. Hooker and G. A.
W. Arnott, Esq.

Eng. Fl. English Flora. Sir J. E. Smith.

Eng.Bot. English Botany. Ditto.

Cyb.Brit. Cybele Britannica. Mr. H. C. Watson.

Phyt. Phytologist.

Ips. Fl. Ipswich Flora. Loudon's Mag. Nat. Hist.
Vol. 3.

Hist. B. History of Bury, by Gillingwater. List
of Plants by Sir T. G. Cullum.

Hist. F. History of Framlingham by Loder. List of
Plants by the Poet Crabbe.

Hist. Y. History of Yarmouth, by Messrs. Paget,
Brothers. History of Plants growing
about Yarmouth.

* An Asterisk prefixed to a species indicates that it is
naturalized.

ADDENDA.

Page 2. Add to the localities of Myosurus minimus :
Fields at Little Welnetham. Fields and gardens at Herring-fleet, plentiful.

Page 13. Add to localities of Holosteum umbellatum :
Old walls at Eye, plentiful.

Page 51. After Hieracium umbellatum, add :—

6. H. *subaudum*, Savoy H.

Var. boreale, Farnham, Bucklesham, Otley and Bentley.

Page 105, Add :—

CETERACH *officinarum*, Willd. On a wall at Lavenham.

Page 106. After TRICHOSTOMUM *lanuginosum* add :—

T. *canescens*, Hedw. Sandy beach about Yarmouth and Lowestoft.

Page 107, line 38 :—

For " Suffolk " read " Santon Downham."

Page 109. Add :—

JUNGERMANNIA *excisa*, Dicks. Hedgebanks and heathy places about Yarmouth.

—— *incisa*, Schrad. Bogs at Herringfleet and Westleton.

—— *resupinata*, L. Under the trailing stems of Ericæ, in shady places.

—— *heterophylla*, Schrad. On decaying stumps, especially of Alder.

—— *anomala*, Hook. Bogs at Westleton.

—— *Francisci*, For " Suffolk " read " Moist ground near Herringfleet Decoy."

—— *pinguis*, For " Suffolk " read " Herringfleet."

To the localities of JUNG. *crenulata, connivens, exsecta* and *setacea*, ADD Bogs at Westleton.

CORRIGENDA.

Page 22. line 2. For " Freston " *read* " Friston."

„ 42. last line. Ditto Ditto.

THE FLORA

OF

SUFFOLK.

Bury St. Edmund's:
Printed by W. T Jackson, Bookseller, etc., Angel Hill.
MDCCCLX.

CLASS I.

DICOTYLEDONES.—DICOTYLEDONS.

DIV. I. ANGIOSPERMÆ.—ANGIOSPERMS.

SUB-CLASS I. THALAMIFLORÆ.— THALAMI-FLORES.

ORDER I. RANUNCULACEÆ. — RANUNCU-LANTHS, THE RANUNCULUS FAMILY.

I. CLEMATIS. Clematis.

1. C. *vitalba*, L. Common C. (Traveller's Joy.) June—July.

Common in hedges and thickets.

II. THALICTRUM. Thalictrum.

2. T. *minus*, L. Lesser T. June—July.

Rare. Hedges in Shaker's Lane, Hospital Road, and near Cemetery, Bury. Stowlangtoft, Higham & Mildenhall.

3. T. *flavum*, L. Yellow T. (Meadow Rue.) June —July.

Not very general Watery places, at Somerleyton, Bungay, Great Glemham, Stratford St. Andrew's, Bergholt, Monk's Eleigh and Mildenhall. Near Ipswich. W.L.N. Framlingham. Hist. F.

III. ANEMONE. Anemone.

1. A. *Pulsatilla*, L. Pasque A. April—May.

Rare. Newmarket Heath. Near the Flint Works, Ick-lingham, (extinct ?). On a chalk bank at Cavenham Severals. O.B.G.

B

2. A. *nemorosa*, L. Wood A. April—June.

Frequent in woods and shady places.

IV. ADONIS. Adonis.

1. A.* *autumnalis*, L. Common A. (Pheasant's Eye.) May—July.

Cultivated fields, Hitcham, occasionally. Bury. N.B.G.

V. MYOSURUS. Mousetail.

1. M. *minimus*, L. Common M. April—June.

Ickworth Park, in bare spots, occasionally. Flatford Mill, and about the banks of the Stour. Damp fields at Blundeston and Parham. O.B G. Bungay. N.B.G. Near Gipping. Ips. Fl. Near Browston Hall. Hist. Y.

VI. RANUNCULUS. Ranunculus.

1. R. *aquatilis*, L. Water R. June—August.

 a. *fluitans*, floating W. R. Abundant in the Lark.

 b. *circinnatus*, capillary W. R. Abundant in ditches.

 c. *aquatilis*, common W. R. Abundant in ditches.

 d. *hederaceus*, Ivy-leaved W. R. Lowestoft, near the Old China Works. Snape Bog. Bergholt.

2. R. *Lingua*, L. Great R. (Greater Spearwort.) July—September.

Watery places, rare. Needham Bricet Wood. Mildenhall. Framlingham. Hist. F. Bungay, Beccles Common, Timworth and Cavenham. O.B.G.

3. R. *flammula*, L. Spear R. (Lesser Spearwort.) June—August.

Wet pastures and sides of ditches, common.

5. R. *Ficaria*, L. Figwort R. (Lesser Celandine.) March—May.

Abundant in damp meadows and shady places.

6. R. *sceleratus*, L. Celery-leaved R. April—September.

Not unfrequent by the sides of ditches, &c.

7. R. *auricomus*, L. Wood R. (Goldilocks.) April
—May.
Common in woods and shady situations.

8. R. *acris*, L. Meadow R. (Crowfoot. Buttercups.) June—July.
Meadows and pastures, abundant.

9. R. *repens*, L. Creeping R. May—August.
Common in pastures and wet places.

10. R. *bulbosus*, L. Bulbous R. May—June,
Meadows and pastures, abundant.

11. R. *philonotis*, Ehrh. Hairy R. (R. hirsutus.)
June—September.
Fields and waste ground in a clayey soil. Frequent at
Hitcham, the Bradfields, Bungay, &c.

12. R. *parviflorus*, L. Small-flowered R. May—
August.
Very rare. Hadleigh and Bergholt.

13. R. *arvensis*, L. Corn R. May—July.
Frequent in corn fields. Bury, Hitcham, Otley, Great
Glemham, &c.

VII. CALTHA. Caltha.

1. C. *palustris*, L. Marsh C. (Marsh Marigold.)
March—June.
Common in marshy places.

IX. HELLEBORUS. Hellebore.

1. H. *viridis*, L. Green H. March—April.
Hitcham, Felsham, Benhall and Kettlebaston. Near
Stradbroke, O.B.G. Bury. N.B.G.

2. H. *fœtidus*, L. Fetid H. (Bear's-foot.)
February—April.
About Ickworth and Hitcham. Roadside near Haughley
Station. Cransford, Bungay and Laxfield. O.B.G.

X. AQUILEGIA. Columbine.

1. A* *vulgaris*, L Common C. May—July.
Rare. Little Bricet, Hitcham. Lane between Sicklesmere and Hawstead. Near Bungay.

XI. DELPHINIUM. Larkspur.

1. D* *consolida*, L. Field L. June—July.

Occasionally in corn fields at Barton and east of the Gaol, Bury. About Tuddenham and Hitcham. Brandon. N.B.G. Hall Farm, Aldborough. O.B.G.

ORD. II. BERBERIDACEÆ.—BERBERIDANTHS, THE BARBERRY FAMILY.

I. BERBERIS. Barberry.

1. B. *vulgaris*, L. Common B. May—June.

Frequent in hedges and thickets.

ORD. III. NYMPHÆACEÆ.—NYMPHÆANTHS, THE WATER LILY FAMILY.

I. NYMPHÆA. Nymphæa.

1. N. *alba*, L. White N, (White Water-Lily.) July.

Frequent in ponds, but generally planted. Ditches about Bungay. Great Finborough. Aldborough. Var. minor. Cornard Mere. Cyb. Brit.

II. NUPHAR. Nuphar.

1. N. *lutea*, Sm. Yellow N. (Yellow Water-Lily.) July.

Frequent in streams and ditches.

ORD. IV. PAPAVERACÆ.—PAPAVERANTHS, THE POPPY FAMILY.

I. PAPAVER. Poppy.

2. P. *Rhœas*, L. Field P. June—August.

Abundant in waste and cultivated ground.

3. P. *dubium*, L. Long-headed P. May—July.

In the same situations as the last, but not so common.

4. P. *hybridum*, L. Rough P. May—July.

Sparingly—about Lowestoft, Felixstow, Bury, Saxham and Mildenhall. Near Bradwell Mill. Hist. Y.

5. P. *Argemone*, L. Pale P. May—July.

Frequent in fields, on roadsides, &c.

III. CHELIDONIUM. Celandine.

1. C. *majus*, L. Common C. May—August.

Waste places about villages and ruins, not unfrequent.

V. GLAUCIUM. Glaucium.

1. G. *luteum*, Scop Yellow G. (Horned-Poppy.) June—October.

Scattered along the coast, but seldom plentiful. Lowestoft, Aldborough, Dunwich, Felixstow. &c.

ORD. V. FUMARIACEÆ.—FUMARIANTHS, THE FUMITORY FAMILY.

I. FUMARIA. Fumitory.

1. F. *officinalis*, L. Common F. May—Sept.

a. *capreolata*, rampant F. Hedges, Felixstow.

b. *officinalis*, common F. Abundant in fields, hedgebanks, &c.

d. *parviflora*, small-flowered F. About Mildenhall and Newmarket. O.B.G.

II. CORYDALIS. Corydal.

1. C* *lutea*, D.C. Yellow C. May—August.

Established in many places on old walls, but, doubtless, escaped from gardens.

ORD. VI. CRUCIFERÆ.—CRUCIFERS OR BRASSICANTHS, THE CRUCIFER FAMILY.

II. CHEIRANTHUS. Wallflower.

1. C.* *Cheiri*, L. Common W. (Gilliflower.) May—June.

Ruins and old walls. Bury, Framlingham, Mettingham, Bungay, &c.

B 2

III. BARBAREA. Wintercress.

1. B. *vulgaris,* Br. Common W. (Yellow Rocket.) May—August.

   Common in moist hedges and by the sides of streams and ditches.

IV. NASTURTIUM. Watercress.

1. N. *officinale,* Br. Common W. May—Oct.

   Abundant in brooks and springy places.

2. N. *sylvestre,* Br. Creeping W. June—August.

   Near the margin of Rougham Canal. Near the Docks, Lowestoft. About Framlingham. O.B.G. Bungay. Eng. Fl.

3. N. *palustre,* D.C. Marsh W. (N. terrestre.) June—October.

   More general than the last and in similar situations. Barton Mere, Hitcham, Haughley, Benhall, &c.

4. N. *amphibium,* Br. Great W. (Yellow Cress.) (Armoracia). June—September.

   Side of Coal River, Fornham St. Martin. Framlingham, Bungay.

V. ARABIS. Rockcress.

1. A. *perfoliata,* Lam. Glabrous R. (Tower-Mustard.) (Turritis glabra.) May—July.

   Roadsides and hedge banks, about Rougham, Troston, Great Glemham, Farnham, Brantham and Cosford Union. Belstead. W.L.N. Near Dunwich and Flixton. O.B.G.

3. A. *hirsuta,* Br. Hairy R. June—August.

   Frequent about Bury, Thurston, West Stow, Tuddenham and Mildenhall. Brandon and Sudbourn. O.B.G.

5. A. *Thaliana,* L Thale R. (Wallcress.) (Sisymbrium). May—July.

   Hedge banks at Fornham, Nowton, Hitcham, Great Glemham and Mildenhall.

VI. CARDAMINE. Bittercress.

1. C. *amara,* L. Large B. April—June.

River bank at Fornham St. Martin. Lackford and Stow-market. Kersey mill. About Beccles, Bungay, Weybread and Ufford. O.B.G.

2. C. *pratensis*, L. Meadow B. (Cuckoo-flower.) April—June.

Moist meadows, abundant.

3. C. *impatiens*, L. Narrow-leaved B. May—August.

A few plants appear every year in a shady damp corner of the Botanic Gardens, Bury.

4. C. *hirsuta*, L. Hairy B.

Plentiful in damp and shady situations.

IX. SISYMBRIUM. Sisymbrium.

1. S. *officinale*, Scop. Common S. (Hedge-Mustard.) June—August.

Abundant in waste places, roadsides, &c.

3. S. *Sophia*, L. Fine-leaved S. (Flixweed.) June—August.

Frequent in waste places, about Bury, Mildenhall, Semer, Great Glemham, Blaxhall, &c.

4. S.* *polyceratium*, L. Many podded S. July—August.

Plentiful on a bank near Cannon place, and in School Hall Lane, Bury.

X. ALLIARIA. Alliaria.

1. A. *officinalis*, D.C. Common A. (Garlic-Mustard.) May—June.

Very common everywhere on hedgebanks, &c.

XI. ERYSIMUM. Erysimum.

1. E. *cheiranthoides*, L. Common E. (Treacle-Mustard.) June—August.

Not unfrequent in fields and waste places.

XII. BRASSICA. Brassica.

1. B. *tenuifolia*, Boiss. Wall B. (Diplotaxis.) June—September.

Ruins at Dunwich. Near St. Olave's Bridge.

2. B. *muralis*, Boiss. Sand B. (Diplotaxis.) August—September.

Plentiful in Botanic Gardens, and in chalk pit near Hospital, Bury. Aldborough. Ipswich Docks. Bungay, Cyb. Brit.

5. B. *campestris*, L. Field B. June—July.

Said to grow on the Suffolk coast in the O.B.G.

6. B. *alba*, Boiss. White Mustard. (Sinapis.) June—July.

About Bury, Hitcham, &c.

7. B. *Sinapistrum*, Boiss. Charlock B. (Sinapis arvensis.) May—August

The pest of corn fields in many localities.

8. B. *nigra*, Boiss. Black B. (Black Mustard.) (Sinapis.) June—September.

Hedges at Bury, Great Glemham, Bergholt, Felixstow, &c.

XIII. COCHLEARIA. Cochlearia.

1. C.* *Armoracia*, L. Horseradish C. May.

Naturalized in a few localities, but not indigenous.

2. C. *officinalis*, L. Scurvy C. (Scurvy-Grass.) May—August.

Salt marshes, common.

XIV. ALYSSUM. Alyssum.

2. A. *maritimum*, L. Sweet A. (Koniga.) Aug. —September.

Near Landguard Fort. W.L.N. (An escape.?)

XV. DRABA. Draba

5. D. *verna*, L. Common D. (Whitlow-Grass.) March—May.

Abundant on wall tops, dry banks, &c., in early spring.

XVI. CAMELINA. Camelina.

1. C. * *sativa*, Crantz. Common C. (Gold-of-Pleasure.) June—July.

Waste ground, Lowestoft Docks. Lakenheath field by Wangford. O.B.G.

XVIII. THLASPI. Penny-Cress.

1. *T. arvense*, L. Field P. (Mithridate-Mustard.) May—July.

Plentiful in cultivated and waste ground on the south side of Lake Lothing, near the town of Lowestoft. Bungay. Snape. Freston. W.L.N. Saxmundham, Haughley and Badwell Ash. O.B.G. Framlingham. Hist. F.

XIX. TEESDALIA. Teesdalia.

1. *T. nudicaulis*, Br. Common T. April—June.

Tolerably common throughout the county—more so near the coast, and in the sandy ground in the North West.

XXII. CAPSELLA. Capsell.

1. *C. Bursa-pastoris*, D.C. Shepherd's Purse C. March—September.

Abundant everywhere.

XXIII. LEPIDIUM. Cress.

1. *L. campestre*, Br. Field C. May—August.

Fields and hedgebanks about Lowestoft. Great Glemham, Bungay, Hitcham, &c.

2. *L. Smithii*, H. Smith's C. April—August.

Rare. About Lowestoft. Browston. Eng. Fl.

3. *L.* Draba*, L. Hoary C. May—June.

Plentiful in 1859, in fields and gardens on both sides of Risbygate road, Bury, but doubtless originally imported with seed.

4. *L. latifolium*, L. Broad-leaved C. July—August.

Snape, by the river side. Banks of Orwell. Ips. Fl. Blyborough. O.B.G.

5. *L. ruderale*, L. Narrow-leaved C. May—June.

Waste ground about Ipswich, Aldborough, Yarmouth, Walton Ferry, &c.

XXIV. SENEBIERA. Senebiera.

1. *S. coronopus*, Poir. Common S. (Swine's-Cress.) June—September.

Common in waste ground.

XXVI. CAKILE. Cakile.

1. C. *maritima*, Scop.   Sea C.   (Sea Rocket.)
June—July.

Common on the sandy coast at Lowestoft, Felixstow, Southwold, &c.

XXVII. CRAMBE. Crambe.

1. C. *maritima*, L.   Seakale C.   June.

Rare. Dunwich beach. O.B.G. Southwold. Hist. S.

XXVIII. RAPHANUS. Radish.

1. R. *Raphanistrum*, L.   Wild R.   June—September.

Frequent in cultivated ground throughout the county.

R. *maritimus*, Between Thorpe and Aldborough.

ORD. VII.   RESEDACEÆ.—RESEDANTHS, THE MIGNIONETTE FAMILY.

I. RESEDA. Mignionette.

1. R. *luteola*, L.   Dyer's M.   (Weld.)   June—August.

Common in waste places.

2. R. *lutea*, L.   Cut-leaved M.   June—August.

As common as the last and in similar situations.

ORD. VIII.   CISTACEÆ—CISTANTHS, THE CISTUS FAMILY.

I. HELIANTHEMUM. Rockcist.

1. H. *vulgare*, Gærtn.   Common R.   (Rock-rose.)
July—September.

Plentiful at Saxham, Tuddenham, Rougham, &c., but apparently wanting in some districts.

ORD. IX. VIOLACEÆ.—VIOLANTHS, THE
VIOLET FAMILY.

I. VIOLA. Violet.

2. V. *odorata*, L. Sweet V. March—April.

Generally common, and in some districts abundant.
Occurs purple, pinkish, and often white.

2. V. *hirta*, L. Hairy V. April—May.

Hedge banks. Hitcham, Hawstead, Chevington, Ickworth,
&c.

4. V. *canina*, L. Dog V. April—August.

a. *flavicornis*, Dwarf D.V. Lowestoft and Bungay.

b. *canina*, Common D.V. Very general.

5. V. *tricolor*, L. Pansy V. (Heartsease.) April—
August.

Common in fields and gardens.

ORD. X. FRANKENIACEÆ.—FRANKENIANTHS,
THE FRANKENIA FAMILY.

I. FRANKENIA. Frankenia.

1. F. *lævis*, L. Common F. (Sea-heath.)
July—August.

Aldborough mere, Walberswick, and in salt marshes near
Braydon.

ORD. XI. CARYOPHYLLACEÆ.—SILENANTHS,
THE PINK FAMILY.

1. DIANTHUS. Pink.

2. D. *armeria*, L. Deptford P. July—August.

Sparingly on a hedge bank at Barton. About Sudbury and
Bungay.

3. D. *deltoides*, L. Maiden P. June—September.

West Stow. Icklingham, Tuddenham and Culford heaths.

II. SAPONARIA. Saponaria.

1. S.* *officinalis*, L. Common S. (Soapwort.)
July—August.

About Barningham, Marlesford, Kettlebaston, Blakenham, Bramford and Bungay. Westleton and Beccles. O.B.G.

III. SILENE. Silene.

2. S. *inflata*, Sm. Bladder S. (Bladder Campion.) June—August.

Common everywhere.

Var. maritima.—Aldborough, Dunwich and Landguard Fort.

3. S. *otites*, Sm. Spanish S. June—August.

West Stow, Icklingham and Risby heaths. Tuddenham Churchyard. Mildenhall and Thetford Newmarket. N.B.G.

5. S. *gallica*, L. Small-flowered S. (S. anglica.) June—October.

Fields about Westley, Higham, Tuddenham, Lowestoft and Great Glemham. Farnham, Woodbridge, Sutton, Lound and Bradwell.

Var. quinquevulnera, has been found at Sutton.

6. S. *conica*, L. Striated S. May—July.

Rare. In sandy fields, about Icklingham, Mildenhall and Brandon.

7. S. *noctiflora*, L. Night S. July—August.

Not uncommon in corn fields about Bury, Rougham, Hitcham, Great Glemham, &c.

IV. LYCHNIS. Lychnis.

1. L. *vespertina*, Sibth. White L. June—Sept.

Common in fields and waste places.

2. L. *diurna*, Sibth. Red L. June—July.

Not quite so frequent as the last, and usually in damp hedges.

3. L. *Githago*, Lam. Com. L. (Corn Cockle.) (*Agrostemma.*) June—August.

Common in corn fields.

4. L. *Flos-cuculi*, L. Meadow L. (Ragged Robbin.) May—June.

Abundant in marshes, &c.

V. SAGINA. Pearlwort.

1. S. *procumbens*, L. Procumbent P. May— September.

This plant and its apetalous variety (S. *apetala*) are very generally distributed. *Var. maritima* is recorded from Southwold and Gorleston.

3. S. *nodosa*, Fenzl. Knotted P. (Spergula.) July, August.

Not so common as the former. It has been observed at Benhall, Sizewell, Lowestoft, Bungay, Tuddenham, and Higham.

VII. ARENARIA. Sandwort.

3. A. *tenuifolia*, L. Fine-leaved S. May, June. Haberden, Bury. Livermere, Westley, Higham, Risby heath, Tuddenham, Barton Mills, and Mildenhall.

4. A. *peploides*, L. Ovate S. (Sea Purslane,) (Honchenya.) May—July.

Common on the sandy coast.

5. A. *serpyllifolia*, L. Thyme-leaved S. June— August.

Abundant on walls and in dry places.

7. A. *trinervis*, L. Three-nerved S. June, July.

Common in damp shady places.

VIII. MŒNCHIA. Mœnchia.

1. M. *erecta*, Sm. Upright M. May, June.

West Stow heath. Nacton and Snape. Lowestoft Denes. Bungay. Framlingham. Hist. F.

IX. HOLOSTEUM. Holosteum.

1. H. *umbellatum*, L. Umbellate H. April.

This rare plant formerly grew on thatched roofs just beyond the Railway Station, Northgate, Bury; but it has lately been searched for in vain.

X. CERASTIUM. Cerast.

1. C. *vulgatum*, L. Common C. (Mouse-Ear Chickweed.) March—September.

a. b. & c. Common.

d. *tetrandrum.* Lowestoft Denes. Orford Beach. Near Yarmouth. N.B.G. Yoxford. O.B.G. Framlingham. Hist. F.

2. C. *arvense*, L. Field C. April—August.

Not generally very common, but plentiful about Bury, Mildenhall, and the sandy ground in the North West.

XI. STELLARIA. Starwort.

1. S. *aquatica*, Scop. Water S. (Malachium.) (Water Mouse-Ear Chickweed.) July, August.

Frequent about Bury, Hardwick, Lackford, Trimley, Bergholt, Great Glemham, and Badingham.

3. L. *media*, L. Chickweed S. Jan—Dec.

Abundant everywhere.

4. S. *uliginosa*, Murr. Bog. S. May, June.

Moist ground at Bury, West Stow, Snape, Benhall, Bungay, &c.

5. S. *graminea*, L. Lesser S. (Lesser Stitchwort.) May—August.

Common among low bushes, &c.

6. S *glauca*, With. Glaucous S. May—July.

Sides of ditches and wet places at West Stow, Icklingham, and Cavenham. Mildenhall, Burgh Common, and Bradwell. O.B.G.

7. S. *holostea*, L. Great S. (Greater Stitchwort,) (All-bones.) April—June.

Common in hedges.

XII. SPERGULARIA. Sand Spurry.

1. S. *rubra*, Pers. Common S. *(Arenaria rubra et marina.)* June—September.

Both varieties common near the coast; the smaller with pink flowers occurs at Bungay, Troston, Tuddenham, &c.

XIII. SPERGULA. Spurry.

1. S. *arvensis*, L. Corn S. June—August.

Corn-fields and waste places in a light soil. Abnndant about Lowestoft, Tuddenham, &c.

ORD. XIII. TAMARISCINEÆ.—TAMARICANTHS,
THE TAMARISC FAMILY.

I. TAMARIX. Tamarisc.

1. T.* *gallica*, L. Common T. July.
Naturalized about Landguard Fort, &c.

ORD. XIV. HYPERICINEÆ.—HYPERICANTHS,
THE HYPERICUM FAMILY.

I. HYPERICUM. Hypericum.

1. H.* *calycinum*, L. Large-flowered H. July
—September.
Woods on Nacton heath, plentiful, and apparently wild.

2. H. *Androscemum*, L. Tutsan H. June—Aug.
Hedges at Holton and Weston. Mr. Wigg.

3. H. *perforatum*, L. Common H. (St. John's
Wort.) July—September.
Very general in hedges, woods, &c.

4. H. *dubium*, Leers. Imperforate H. July, Aug.
About Offton. In a lane near Framlingham. O.B.G.

5. H. *quadrangulum*, L. Square-stalked H. July.
Common in wet places.

6. H. *humifusum*, L. Trailing H. July.
Rather local. Rougham and Risby heaths. Bergholt
and Blaxhall. Otley wood.

8. H. *pulchrum*, L. Slender H. June, July.
Frequent in dry or heathy situations.

9. H. *hirsutum*, L. Hairy H. July, August.
Not uncommon throughout the county, except perhaps
near the sea and in the north-west sandy ground.

11. H. *Elodes*, L. Marsh H. July, August.
Bogs at Tuddenham. Belton. N.B.G. About Yarmouth.
Hist. Y.

ORD. XV. LINACEÆ.—LINANTHS, THE FLAX FAMILY.

I. LINUM. Flax.

1. L.* *usitatissimum*, L. Common F. July.

Occasionally in various places. Hitcham, Lowestoft, and Ickworth park.

2. L. *perenne*, L. Perennial F. June, July.

Rare. Ixworth and Stowlangtoft.

3. L. *angustifolium*, Huds. Pale F. May—Sept.

Rare. Darsham. Eng. Bot.

4. L. *catharticum*, L. Cathartic F. June—Sept.

Common in all districts.

II. RADIOLA. Allseed.

1. R. *millegrana*, Sm. Common A. July, Aug.

Lowestoft Denes and on a small heath south of Lake Lothing. Somerleyton, Herringfleet, and Tuddenham heaths. —Belton. Hist. Y.—Culford and Timworth. Hist. Bury.

ORD. XVI. MALVACEÆ. MALVANTHS, THE MALLOW FAMILY.

II. MALVA. Mallow.

1. M. *rotundifolia*, L. Dwarf M. June—Sept.

Common by roadsides, &c.

2. M. *sylvestris*, L. Common M. June—Sep.

Very common in hedges.

3. M. *moschata*, L. Musk M. July, August.

Local. Rattlesden, Bergholt, Sudbury, Wherstead, Belstead, and Bungay. Martlesham. J.W.G.

III. ALTHÆA. Althæa.

1. A. *officinalis*, L. Marsh A. (Marsh-Mallow.) August, September.

Plentiful about Somerleyton & Haddiscoe. Snape marshes. Walton ferry. Between Dunwich and Sizewell. J.W.G.

ORD. XVII. TILIACEÆ.—TILIANTHS, THE LIME FAMILY.

I. TILIA. Lime.

1. T.* *europæa*, L. Common L. July.
Naturalized in some districts.

ORD. XVIII. GERANIACEÆ.—GERANIANTHS, THE GERANIUM FAMILY.

I. GERANIUM. Geranium. (Crânesbill.)

1. G. *sanguineum*, L. Blood G. July.
On the left-hand of the road from Bury to Ixworth, opposite Barton hall. Westley. Bergholt. Near Woodbridge. O.B.G.

2. G. *phæum*, L. Dusky G. May, June.
Bergholt and Ellingham. On the borders of a wood at Ampton, probably escaped. On a bank at Ashbocking. O.B.G.

4. G. *pratense*, L. Meadow G. June—Sept.
In a meadow at Darsham. O.B.G. "Kentford, plentiful."

5. G. *pyrenaicum*, L. Mountain G. June, July.
About Bury, Great Glemham, Bungay, Saxmundham, and Cosford Union —Beccles. O.B.G.

6. G. *Robertianum*, L. Herb-Robert G. May—September.
Common everywhere. *Var. purpureum* at Aldborough.

7. G. *lucidum*, L. Shining G. May—August.
Rather local, but plentiful about Bury, Bergholt, Barham, &c.

8. G. *molle*, L. Dove's-foot G. April—August.
Abundant everywhere.

9. G. *pusillum*, L. Small-flowered G. June—September.
Plentiful in all districts.

c 2

10   G. *rotundifolium*, L.   Round-leaved G.   June,
July.

Rare.   Stoke hill, Ipswich.—Westleton.   J.W.G.

11.   G. *dissectum*, L.   Cut-leaved G.   May—Aug.

Common in fields and waste places.

12.   G. *columbinum*, L.   Long-stalked G.   June,
July.

Very scarce.   About Bungay.

II.   ERODIUM.   Erodium.   (Storksbill.)

1.   E. *cicutarium*, L'Her.   Common E.   June—
September.

Abundant nearly everywhere.

2.   E. *moschatum*, L'Her.   Musk E.   June, July.

Rare.   Sternfield.   Bradwell.   O.B.G.

III.   OXALIS.   Oxalis.

1.   O. *acetosella*, L.   Sorrel O.   (Wood-Sorrel.)
May.

Not uncommon in woods and shady places.   Rushbrooke,
Hitcham, Great Glemham, Lowestoft, &c.

ORD.   XX.   POLYGALACEÆ.—POLYGALANTHS,
THE MILKWORT FAMILY.

I.   POLYGALA.   Milkwort.

1.   P. *vulgaris*, L.   Common M.   May—Sep.

Frequent, especially in the neighbourhood of heaths.

ORD.   XXI.   ACERACEÆ.—ACERANTHS,   THE
MAPLE FAMILY.

I.   ACER.   Maple.

1.   A. *campestre*, L.   Common M.   May, June.

Woods and thickets, frequent.

2. A.* *Pseudo-platanus*, L. Sycamore M. May, June.

Naturalized in a few localities.

SUB-CLASS II. CALYCIFLORÆ.—CALYCIFLORES.
(Sect. I. Polypetalous.)

ORD. XXII. CELASTRACEÆ.—CELASTRANTHS, THE CELASTRUS FAMILY.

I. EVONYMUS. Spindle-Tree.

1. E. *europæus*, L. Common S. May, June.

Frequent about Bury, Hitcham, Great Glemham, Bungay, &c. Plentiful in the Hyde, Westley.

ORD. XXIII. RHAMNACEÆ.—RHAMNANTHS, THE BUCKTHORN FAMILY.

I. RHAMNUS. Buckthorn.

1. R. *catharticus*, L. Common B. May—July.

Not uncommon. Rushbrook, Barton, Hitcham, Offton, and Bungay. About Framlingham, Woolpit wood, and Hoxne. O.B.G.

2. R. *Frangula*, L. Alder B. May, June.

Rather scarce. Hitcham. Otley wood. About Bungay.

ORD. XXIV. PAPILIONACEÆ.—LEGUMENS OR VICIANTHS, THE PEAFLOWER TRIBE.

I. ULEX. Furze.

1. U. *europæus*, L. Common F. (Gorse. Whin.) February—July.

Abundant on heaths and sandy ground.

2. U. *nanus*, Forst. Dwarf F. July—November.

Not so plentiful as the common F., but very general on our heaths.

II. GENISTA. Genista.

1. G. *tinctoria*, L. Dyer's G. (Greenweed.)
July, August.

Waste ground at Hardwick, Rougham, Great Glemham,
Bungay, &c.

2. G. *pilosa*, L. Hairy G. May—September.

Rare. Tuddenham heath. Culford, Cavenham, and Lackford. O.B.G.

3. G. *anglica*, L. Needle G. (Petty Whin.)
May, June.

West Stow & Tuddenham heaths. Common about Bungay.

III. SAROTHAMNUS. Broom.

1. S. *scoparius*, Wimm. Common B. April—
June.

Common almost everywhere.

IV. ONONIS. Ononis.

1. O. *arvensis*, L. Restharrow O. June—Sep.
Both the armed and unarmed varieties are general throughout the county.

V. MEDICAGO. Medick.

1. M. *falcata*, L. Sickle M. June, July.

Frequent in hedges about Bury, Thurston, Stowlangtoft,
Brandon, Lakenheath, and Mildenhall. Old Churchyard, Dunwich. Sudbourn and Orford. O.B.G

3. M. *lupulina*, L. Black M. (Nonsuch.) May
—August.

Common in dry and gravelly places.

4. M. *denticulata*, Willd. Denticulate M. April
—August.

Rare. Orford beach.

5. M. *maculata*, Willd. Spotted M. May—Aug.

Not uncommon. Haberden, Bury. Hitcham, Bungay,
Ipswich, Stowlangtoft, Orford, Saxmundham, &c.

6. M. *minima*, L. Bur M. May—July.

Local, but plentiful in a field near lcklingham heath. Risby heath, Mildenhall, Fakenham, Felixstow, Bawdsey ferry, and near Orford castle.

VI. MELILOTUS. Melilot.

1. M. *officinalis*, L. Common M. June—Aug.

Common about Bury, Thurston, Elmswell, Great Glemham, Hitcham, &c.

3. M. *alba*, Lam. White M. July, Aug.

Rare. About Mildenhall. Waste ground near the Haddiscoe railway station—a few plants in 1858.

VII. TRIGONELLA. Trigonel.

1. T. *ornithopodioides*, D.C. Bird's-foot T. (Fenugreek.) June, July.

Rare. Orford beach. Bawdsey ferry. W.L.N. Aldborough, Lowestoft, and Yarmouth. O.B.G.

VIII. TRIFOLIUM. Clover.

2. T. *arvense*, L. Hare's-foot C. July, Aug.

Abundant in a light sandy soil.

4. T. *ochroleucum*, L. Sulphur C. June—Aug.

Frequent throughout Suffolk, except perhaps near the sea and in the north-west.

5. T. *pratense*, L. Purple C. May—September.

Common in meadows and pastures.

6. T. *medium*, L. Zigzag C. June—September.

About Bury,'Stowlangtoft, Bungay, Great Glemham, &c.

7. T. *maritimum*, Huds. Sea C. June, July.

Salt marshes near Yarmouth. O.B.G.

8. T. *striatum*, L. Knotted C. June, July.

Frequent about Bury. Hitcham, Great Glemham, Blaxhall and Bungay.

10. T. *scabrum*, L. Rough C. May—July.

Plentiful on the sandy coast and in the sandy north-west ground, less common elsewhere. Bungay.

12. T. *glomeratum*, L. Clustered C. June.

Not very general. About Freston, Blaxhall, Bungay, and Great Glemham. Aldborough, Middleton, and Dunwich. O.B.G.

13. T. *suffocatum*, L. Suffocated C. June, July.

Not unfrequent near the sea. Lowestoft, Yarmouth, and Bungay.—Dunwich, Southwold, Aldborough, and Landguard fort. O.B.G.

15. T. *subterraneum*, L. Subterranean C. May, June.

Haberden, Bury. Rougham heath, Stowlangtoft, Great Glemham, Aldborough, and Bungay.

16. T. *fragiferam*, L. Strawberry C. July, Aug.

Not uncommon in most districts, but most general in the neighbourhood of the sea.

17. T. *repens*, L. White C. (Dutch C.) May —September.

Common everywhere.

18. T. *agrarium*, L. Hop C. (T. procumbens.) June—August.

Equally common with the last.

19. T. *procumbens*, L. Lesser C. (T. minus.) June, July.

As frequent as the two foregoing.

20. T. *filiforme*, L. Slender C. June, July.

Rare. Snape, Great Glemham, and Bungay.

IX. LOTUS. Lotus.

1. L. *corniculatus*, L. Common L. (Bird's-foot Trefoil.) June—August.

a. and b. Common. c. Local.

2. L. *angustissimus*, L. Slender L. (L. diffusus.) May—August.

Rare. Below Woodbridge, between the river wall and marsh ditch.

X. ANTHYLLIS. Anthyllis.

1. A. *vulneraria,* L. Common A. (Lady's-Fingers.) June—August.

On the borders of Risby heath. Near Barton mere. Westley.

XI. ASTRAGALUS. Astragal.

1. A. *hypoglottis,* L. Purple A. June, July.

Not common. Risby heath. Mildenhall. Newmarket heath. O.B.G.

3. A. *glycyphyllus,* L. Sweet A. (Milkvetch.) June—September.

Not very general. Fornham St Martin, Pakenham, Barton, Stowlangtoft, Whatfield and Claydon. Coddenham. O.B.G.

XIV. ORNITHOPUS. Bird's-foot.

1. O. *perpusillus,* L. Common B. May—July.

Not unfrequent in a dry barren soil. About Bury, Rougham, West Stow, Snape, Aldborough, &c.

XV. HIPPOCREPIS. Hippocrepis.

1. H. *comosa,* L. Common H. May—August.

Very plentiful near the chalk pit on Risby heath. Newmarket heath. Mildenhall, Offton and Somersham.

XVI. ONOBRYCHIS. Saintfoin.

1. O.* *sativa,* Lam. Common S. June, July.

In various places. Perhaps wild at Mildenhall, and near the coast.

XVII. VICIA. Vetch.

1. V. *hirsuta,* Koch. Hairy V. June—August.

Not uncommon about Ickworth, Bradfield, Great Glemham, Aldborough, Framlingham, &c.

2. V. *tetrasperma,* Mœnch. Slender V. June, August.

Not so general as the last. About Honington, Hitcham, Offton, Rendham, Bungay, &c.

3. V. *Cracca,* L. Tufted V. June—August.

Common in hedges and thickets.

6. V. *sepium*, L. Bush V. June—August.
Equally distributed with the foregoing.

7. V. *lutea*, L. Yellow V. June—August.
Rare. Orford beach. Aldborough. Eng. Fl.

8. V.* *sativa*, L. Common V. May, June.
Common in many places, if it may be considered wild.

9. V. *lathyroides*. L. Spring V. April—June.
Not unfrequent. About Bury, West Stow, Mildenhall, Little Glemham Park and Bungay.—Aldborough and Westleton. O.B.G.

XVIII. LATHYRUS. Pea. (Vetchling.)

1. L. *Nissolia*, L. Grass P. May, June.
Local. About Stowlangtoft, Naughton, Bungay & Ipswich. Felixstow. W.L.N. Darsham and Parbam. O.B.G.

2. L. *Aphaca*, L. Yellow P. May—August.
Rare. Naughton. Clay pit at Bungay. Gravel pit near Sicklesmere. O.B.G.

4. L. *pratensis*, L. Meadow P. July, August.
Common in moist places.

5. L. *sylvestris*, L. Everlasting P. June—Aug.
Plentiful about Gallow hill and Wood Hall Lane, Sudbury. Woods at Great Saxham. O.B.G.

6. L. *palustris*, L. Marsh P. June—August.
Occasionally in marshes near the sea. Oulton marshes, directly south of the Broad. Lakenheath. Belton. Hist. Y. Beccles, Worlingham, Flixton, and Burgh castle. O.B.G.

7. L. *maritimus*, Bigel. Sea P. (Pisum.) July, August.
Rare. Aldborough beach. Orford beach. Ray.

1. L. *macrorrhizus*, Wimm. Tuberous P. (Orobus tuberosus.) May—July.
Rare. About Honington.

ORD. XXV. ROSACEÆ.—ROSANTHS, THE ROSE FAMILY.

I. PRUNUS. Prunus.

1. P. *communis*, Huds. Blackthorn P. (Sloe and Bullace.) April, May.
Abundant in hedges and thickets.
2. P.* *Cerasus*, L. Cherry P. May.
Woods and hedgerows. Bradfield, Hitcham, &c.
3. P. *Padus*, L. Bird-Cherry P. May.
Rare. Woods near Framlingham. O.B.G.

II. SPIRÆA. Spiræa.

1. S. *Ulmaria*, L. Meadow S. (Meadow-Sweet.) June—August.
Common in moist meadows and by streams and ditches.
2. S. *Filipendula*, L. Common S. (Dropwort.) June, July.
Frequent. Barton park. Westley, Culford, Cavenham, Tuddenham, &c.

IV. GEUM. Avens.

1. G. *urbanum*, L. Common A. (Herb-Bennet.) June—August.
Common everywhere.
2. G. *rivale*, L. Water A. May—July.
Not unfrequent. Link woods, Rushbrooke, plentiful, with *intermedium*. Bradfield St. George, Hitcham, Bergholt, and elsewhere.

V. RUBUS. Rubus.

1. R.* *idæus*, L. Raspberry R. June, July.
Woods at Hitcham and Kesgrave.
2. R. *fruticosus*, L. Blackberry R. (Bramble.) July, August.
Abundant in hedges and thickets.
3. R. *cæsius*, L. Dewberry R. June, July.
Moist hedges and sides of ditches, common.

VI. FRAGARIA. Strawberry.

1. F. *vesca*, L. Common S. May—July

Plentiful in woods.

VII. POTENTILLA. Potentil.

1. P. *Fragariastrum*, Ehrh. Strawberry-leaved P. March—May.

Very general on dry banks and in woods.

2. P. *reptans*, L. Creeping P. (Cinquefoil.) June —September.

Common on roadsides, &c.

3. P. *Tormentilla*, Sibth. Tormentil P. June— August.

Abundant in heathy ground. *Var. reptans.*—In a lane at Belton and wood at Corton. O B.G.

4. P. *argentea*, L. Hoary P. June, July.

Not unfrequent in a gravelly soil. About Barton, Fornham, Troston, Woolpit, Tuddenham, Bungay, Blaxhall, Great Glemham, and Polstead.

5. P. *verna*, L. Spring P. April—June.

Rare. Chalk bank at Cavenham Severals. O.B.G.

7. P. *anserina*, L. Goose P. (Silver-weed.) June, July.

Very common on roadsides and in a sandy soil.

9. P. *Comarum*, Nestl. Marsh P. (Comarum palustre.) May—July.

Not very general. Plentiful in bogs at Tuddenham, West Stow, Sudbury and Bungay.—Mildenhall. J. T.

IX. ALCHEMILLA. Alchemil.

3. A. *arvensis*, Scop. Field A. (Parsley-Piert.) May—August.

Plentiful in cultivated and waste ground.

X. SANGUISORBA. Sanguisorb.

1. S. *officinalis*, L. Burnet S. (Great Burnet.) June—August.

Lakenheath. F. K. Eagle, Esq.

XI. POTERIUM. Poterium.

1. P. *Sanguisorba*, L. Burnet P. (Salad-Burnet.) June—August.

Frequent in a chalky soil. Borders of fields, railway embankments, &c.

XII. AGRIMONIA. Agrimony.

1. A. *Eupatoria*, L. Common A. June, July.

Common about roadsides, borders of fields, &c.

XIII. ROSA. Rose.

1 R. *pimpinellifolia*, L. Burnet R. (R. spinosissima.) May.

Very general towards the coast.

2. R.* *villosa*, L. Downy R. (R. tomentosa.) June, July.

Hedges and thickets, not common. About Hitcham, Bungay, Belton Common. Lane between Bradwell and Burgh castle.

3. R. *rubiginosa*, L. Sweetbriar R. June, July.

Abundant in hedges.

4. R. *canina*, L. Dog R. June, July.

As common as the former.

5. R. *arvensis*, L. Field R. June, July.

Hedges and thickets. Somewhat less frequent than the two last species.

XIV. PYRUS. Pyrus.

1. P. *communis*, L. Pear P. April, May.

Hedges about Bungay.—In the neighbourhood of the sea. Hist. Y.

2. P. *Malus*, L. Apple P. (Crab.) May.

Common in hedgerows.

3. P.* *Aria*, Ehrh. Beam P. (White Beam-tree.) May, June.

Near Bungay. O.B.G.

4. P. *torminalis*, Ehrh. Cut-leaved P. (Wild Service-tree.) May, June.

About Bungay.—Darsham. O.B.G.

5. P.* *Aucuparia*, Gærtn. Rowan P. (Mountain-Ash. May, June.

Woods, &c.

XV. CRATÆGUS. Hawthorn.

1. C. *Oxyacantha*, L. Common H. (May. White-thorn.) May, June.

Abundant in old parks, hedges, &c.

ORD. XXVI. ONAGRACEÆ.—ONAGRANTHS, THE ŒNOTHERA FAMILY.

I. EPILOBIUM. Epilobe.

2. E. *hirsutum*, L. Great E. (Great Willow-herb.) July, August.

Plentiful by the sides of rivers and ditches.

3. E. *parviflorum*, Schreb. Hoary E. July, Aug.

With the former, but less common.

4. E. *montanum*, L. Broad E. June, July.

Frequent on dry banks, walls, &c.

5. E. *roseum*, Schreb. Pale E. July, August.

Scarce. Great Glemham.

6. E. *tetragonum*, L. Square E. (E. obscurum.) July, August.

Not uncommon, but probably frequently passed by for allied species. Pakenham Fen, Great Glemham, and Hitcham.

7. E. *palustre*, L. Marsh E. July, August.

Sparingly found in a few localities. Bogs at Tuddenham, Mildenhall, and Hitcham.—Framlingham. Hist. F.

II. ŒNOTHERA. Œnothera. (Onagra. Tournef.)

I. Œ. *biennis*, L. Common Œ. (Evening-Primrose.) July—September.

Woodbridge. Eng. Fl. Plentiful on the sandy cliff at Kirkley till the present pathway was made from Lowestoft to Pakefield. Sandy ground at Sutton.

IV. CIRCÆA. Circæa.

1. C. *lutetiana*, L. Common C. (Enchanter's Nightshade.) June—August.

Frequent in damp shady plcaes.

V. MYRIOPHYLLUM. Myriophyll.

1. M. *spicatum*, L. Spiked M. (Water-Milfoil.) June, July.

Ditches, &c. not uncommon. Somerleyton, Oulton, Otley, Hitcham, Bradfield, and elsewhere.

2. M. *verticillatum*, L. Whorled M. July, Aug.

Perhaps not so frequent as the last. Friars' lane ditches, Bury. St. Olave's bridge & Bungay.—Gorleston. N.B.G. —Middleton, Wenhaston, and Bradwell. O.B.G.

VI. HIPPURIS. Marestail.

1. H. *vulgaris*, L. Common M. June, July.

Plentiful in ditches about Somerleyton, Haddiscoe, Herringfleet, Melford Hall Mill, Bungay, Mildenhall, and Fornham.—Framlingham. Hist. F.—About Ipswich. W.L.N.

ORD. XXVII. LYTHRACEÆ.—LYTHRANTHS, THE LYTHRUM FAMILY.

I. LYTHRUM. Lythrum.

1. L. *salicaria*, L. Spiked L. (Purple-Loosestrife.) July—September.

Very common by the sides of rivers and ditches.

2. L. *hyssopifolium*, L. Hyssop L. June—Oct.

Very rare. Fields near Barton Mere. O.B.G.

II. PEPLIS. Peplis.

1. P. *portula*, L. Common P. (Water-Purslane.) July, August.

Probably not uncommon, though few localities are recorded. Watery places, North Denes, near the lighthouse, Lowestoft. Margins of bogs on Tuddenham heath, Snape and Bungay.

ORD. XXVIII.   CUCURBITACEÆ.—CUCURBI-TANTHS,   THE GOURD FAMILY.

I. BRYONIA.   Bryonia.

1. B. *dioica*, L.   Common B.   May—September.

Very common in hedges, &c.

ORD. XXIX. PORTULACEÆ.—PORTULACANTHS, THE PURSLANE FAMILY.

I. MONTIA.   Montia.

1. M. *fontana*, L.   Water M.   (Blinks.   Water-Chickweed.)   April—August.

Not unfrequent in springy places.   Near the old China Works, Lowestoft.   Boggy ground on Tuddenham heath. St. Olave's Bridge.   Snape.

ORD. XXX.   PARONYCHIACEÆ.—PARONYCHI-ANTHS,   THE PARONYCHIA FAMILY.

II. HERNIARIA.   Herniary.

1. H. *glabra*, L.   Common H.   (Rupture-wort.) July—August.

Rare.   Roadside near Higham station.   Great Barton. Fallows near Barrow Bottom and Henham.   O.B.G.

IV. SCLERANTHUS.   Scleranthus.

1. S. *annuus*. L.   Annual S.   (Knawel.)   July, August.

Not uncommon in a gravelly or sandy soil.

2. S. *perennis*, L.   Perennial S.   June—Sept.

Less frequent than the *annual S.*   Lackford, West Stow, Culford, and Icklingham heaths.—About Elden, Barnham, and Thetford.   O.B.G.

ORD. XXXI. CRASSULACEÆ.—CRASSULANTHS,
THE CRASSULA FAMILY.

I. TILLÆA. Tillæa.

1. T. *muscosa*, L. Mossy T. June, July.

Not uncommon in a sandy, heathy soil, near the sea ; and
in the-north-west, especially near drift-ways or trodden
ground. About Lowestoft, Bungay, Aldborough, Dunwich, and Westleton. Holton, West Stow and Tuddenham heaths. Icklingham, Mildenhall, and Brandon.
Gravel pit near the house of industry, Nacton. Hoxne.
O.B.G.

III. SEDUM. Sedum.

2. S. *Telephium*, L. Orpine S. July, August.

Occasionally in hedge-banks, &c., as at Welnetham,
Hitcham, Bungay, and Polstead. — Framlingham.
Hist. F.

3. S. *anglicum*, Huds. English S. June—Aug.

Near the sea. North Denes, Lowestoft. Aldborough.—
Bawdsey ferry. W.L.N.

4. S. *dasyphyllum*, L. Thick-leaved S. June,
July.

Plentiful on the ruined walls of the Abbey, Bury.

5. S.* *album*, L. White S. July, August.

Roofs of houses and wall tops, Southgate street, Bury.

7. S. *acre*, L. Biting S. (Wallpepper.) June,
July.

Frequent in a sandy or gravelly soil. Bury, Great Glemham, Tuddenham, Mildenhall, Lowestoft, and elsewhere.

9. S. *rupestre*, L. Rock S. July, August.

a. S. *reflexum*, Vine-fields wall, Bury. Bungay, Bergholt, and Stratford.
b. S. *glaucum*, Rough ground near Mildenhall.

IV. SEMPERVIVUM. Housleek.

1. S.* *tectorum*, L. Common H. July.

Not uncommon on the roofs of cottages, &c.

ORD. XXXII. RIBESIACEÆ.—RIBANTHS, THE
RIBES FAMILY.

I. RIBES. Ribes.

1. R.* *Grossularia*, L. Gooseberry R. April,
May.

Thickets and hedges, not unfrequent. Hitcham, Great
Glemham, &c.

2. R.* *rubrum*, L. Currant R. (Red Currant.)
April, May.

Boggy wood at Icklingham. Wood at Santon Downham,
Hitcham, and Great Glemham.

4. R.* *nigrum*, L. Black R. (Black Currant)
April, May.

Great Glemham and elsewhere.

ORD. XXXIII. SAXIFRAGACEÆ.—SAXIFRA-
GANTHS, THE SAXIFRAGE FAMILY.

I. SAXIFRAGA. Saxifrage.

6. S. *granulata*, L. Meadow S. May, June.
Plentiful about Bury, Tuddenham, Little Glemham, Metfield, Ipswich, &c.

9. S. *tridactylytes*, L. (Rue-leaved S.) April—
June.

Common on old walls and barren heathy ground, very con-
spicuous when plentiful owing to the red glands with
which the whole plant is clothed.

II. CHRYSOSPLENIUM. Chrysosplene.

1. C. *oppositifolium*, L. Opposite C. (Golden-
Saxifrage.) April—July.

Apparently a local plant. Mettingham, Shipmeadow,
Hitcham and Brightwell. Freston wood and Downham
Reach wood. W.L.N.

1. C. *alternifolium*, L. Alternate C. April—
June.

About Woodbridge, Great Glemham, and Brightwell.
Plentiful at Mettingham and Shipmeadow.—Alder Car
at Weybread near Needham water-mill. O.B.G.—Freston wood. W.L.N.

III. PARNASSIA. Parnassia.

1. P. *palustris*, L. Marsh P. (Grass-of-Parnassus.) August—October.

Abundant in marshes from Haddiscoe to Oulton. Benhall green, Chillesford, Stowlangtoft, Coney Weston, &c.

IV. DROSERA. Sundew.

1. D. *rotundifolia*, L. Common S. July, Aug.

Common in bogs throughout the county.

2. D. *longifolia*, L. Oblong S. July, August.

Frequent in bogs. Tuddenham, Beccles, Fritton, Lound, &c.

3. D. *Anglica*, Huds. English S. July, Aug.

Bogs near Mildenhall and Lakenheath. Bog on Tuddenham heath with the two former. Probably sometimes passed by for the last species.

ORD. XXXIV. UMBELLIFERÆ.—UMBELLIFERS OR APIANTHS, THE UMBELLATE FAMILY.

I. HYDROCOTYLE. Hydrocotyle.

1. H. *vulgaris*, L. Common H. (Marsh-Pennywort.) White-Rot. May—August.

Frequent in marshes and wet places.

II. SANICULA. Sanicle.

1. S. *europæa*, L. Wood S. June, July.

Not uncommon in damp woods.

IV. ERYNGIUM. Eryngo.

1. E. *maritimum*, L. Sea E. (Sea-Holly.) July, August.

Frequent but not plentiful on the sandy coast.

2. E. *campestre*, L. Field E. July, August.

A few plants were found at Dunwich in 1855 and 1856 by the Rev. E. N. Blomfield.

V. CICUTA. Cowbane.

1. C. *virosa*, L. Water C. (Water-Hemlock.) June—August.

About Brandon, Herringfleet, and Kersey Mill. Fox-hunter's pond, Felsham Hall.—In the Gipping near Ipswich. W.L.N.—Cavenham, Oulton Broad and Dyke, Fritton Broad, Bradwell, and elsewhere about Yarmouth. O.B.G.

VI. APIUM. Apium.

1. A. *graveolens*, L. Celery A. June—Aug.

General throughout the county in fresh and salt marshes and by the side of streams.

VII. HELOSCIADIUM. Helosciad.

1. H. *nodiflorum*, Koch. Procumbent H. July, August.

Abundant in marshes and ditches.

2. H. *inundatum*, Koch. Lesser H. June, July.

Probably not unfrequent, but, from its small size, often escapes observation. Hitcham.

VIII. SISON. Sison.

1. S. *Amomum*, L. Hedge S. (Bastard-Stone-Parsley.) August, September.

Frequent in hedges about Bury, Bradfield, Hitcham, Bungay, Trimley, &c.

IX. PETROSELINUM. Parsley.

2. P. *segetum*, Koch. Corn P. (Sison.) Aug., September.

On a hedge-bank near Westley church.

XI. ÆGOPODIUM. Goutweed.

1. Æ. *Podagraria*, L. Common G. (Bishopweed.) June—August.

Common in moist woods, &c.

XII. CARUM. Carum.

2. C.* *Carui*, L. Caraway C. June.

Pastures about Bergholt.

XIII. SIUM. Sium.

1. S. *latifolium*, L. Broad S. (Water-Parsnip.) July, August.

Ditches about Beccles. Banks of the Waveney between Bungay and St. Olave's. Worlinghám and Cove. About Mildenhall and Brandon.—Marshes between Yarmouth and Burgh castle. O.B.G.—Framlingham. Hist. F.

2. S. *angustifolium*, L. Lesser S. July, August.

Not unfrequent in ditches, &c.

XIV. PIMPINELLA. Pimpinel.

1. P. *Saxifraga*, L. Common P. (Burnet-Saxifrage.) July—September.

Plentiful about Bury, Hitcham, Framlingham, the Glemhams, &c.

2. P. *magna*, L. Greater P. July, August.

Rare. Lane near Hawstead place. Hitcham and Bungay. Woods at Saxham. O.B.G.

XV. BUPLEVRUM. BUPLEVER.

1. B.* *rotundifolium*, L. Hare's-Ear B. (Thorow-Wax.) June, July.

Rare. Cultivated fields at Hitcham and Lakenheath.— Fields at Saxham. O.B.G.

3. B. *tenuissimum*, L. Slender B. Aug., Sept.

Salt marshes at Aldborough. Near the sea at Hollesly. Near Gorleston pier. About Braydon Broad. O.B G.

XVI. ŒNANTHE. Œnanth.

1. Œ. *fistulosa*, L. Common Œ. (Water Dropwort. July—September.

Frequent in marshes and wet places.

2. Œ. *pimpinelloides*, L. Parsley Œ. June—Sept.

a. *pimpinelloides*, Meadow P. Marshes near Oulton Dyke and at Lakenheath.
b. *Lachenalii*, Marsh P. About Mildenhall, Catwade marshes, Snape, and Sizewell.—Near Yarmouth. Hist. Y.

3. Œ. *crocata*, L. Hemlock Œ. July.

Rare. West Stow.—About Ipswich. Ips. Fl.

36

4. Œ. *Phellandrium*, Lam. Fine-leaved Œ. July
—September.

Not very common. Barton mere. Pond in Hitcham wood. Bungay, Great Glemham, and Mildenhall.

XVII. ÆTHUSA. Æthusa.

1. Æ. *Cynapium*, L. Common Æ. (Fool's-Parsley.) July, August.

Plentiful in gardens and cultivated ground.

XVIII. FŒNICULUM. Fennel.

1. F. *vulgare*, Gærtn. Common F. July, Aug.

Frequent near the sea. Lowestoft, Aldborough, Yoxford, Bungay, Burgh Castle, and Fritton. Ampton and Sapiston.

XXI. SILAUS. Silaus.

1. S. *pratensis*, Bess. Meadow S. (Pepper-Saxifrage.) June—September.

Borders of fields and pastures, chiefly in a clayey soil.

XXIV. ANGELICA. Angelica.

1. A. *sylvestris*, L. Wild A. July, August.

Not unfrequent in moist hedges and marshes.

XXV. PEUCEDANUM. Peucedan.

2. P. *palustre*, Mænch. Marsh P. (Hog's-Fennel. Milk-Parsley.) July, August.

Belton bog, Worlingham, Fritton, and Blundeston. O.B.G.

XXVI. PASTINACA. Parsnip.

1. P. *sativa*, L. Common P. July, August.

Common on the borders of fields, &c.

XXVII. HERACLEUM. Heracleum.

1. H. *Sphondylium*, L. Common H. (Cow-Parsnip. Hogs'-weed.) July.

Abundant everywhere.

XXIX. SCANDIX. Scandix.

1. S. *Pecten*, L. Needle S. (Shepherd's-Needle. Venus's-Comb.) June—September.

Common in fields and waste places.

XXXI. BUNIUM. Bunium.

1. B. *flexuosum*, With. Tuberous B. (Earth-Nut. Pig-Nut.)

Plentiful in woods and damp pastures.

XXXII. CHÆROPHYLLUM. Chervil.

1. C. *temulum*, L. Rough C. June, July.

Common in hedges, on roadsides, &c.

2. C. *sylvestre*, L. Wild C. (Anthriscus.) April —June.

Abundant under hedges, borders of fields, &c.

3. C. *Anthriscus*, Lam. Bur C. (Anthriscus vulgaris.) May, June.

Nearly as common as the last, on hedge-banks, &c.

XXXIII. CAUCALIS. Caucalis. (Torilis.)

1. C. *nodosa*, Sm. Knotted C. May—July.

Hedge-banks and under walls, plentiful.

2. C. *Anthriscus*, Huds. Upright C. (Hedge-Parsley.) July—September.

Hedges and waste places, very general.

3. C. *infesta*, Curt. Spreading C. July—Sept.

Not uncommon in fields, &c.

4. C. *daucoides*, L. Small C. June.

Very rare.—Great Saxham. O.B.G.—Newmarket. N.B.G. About Ipswich. Ips. Fl.

XXXIV. DAUCUS. Carrot.

1. D. *Carota*, L. Common C. June—August.

Plentiful on the borders of fields and hedges.

XXXV. CONIUM. Hemlock.

1. C. *maculatum*, L. Common H. June, July.

Frequent in hedges, &c., but not usually abundant.

XXXVII. SMYRNIUM. Smyrnium.

1. S. *olusatrum*, L. Common S. (Alexanders.) April—June.

E

Frequent and plentiful near the sea, and about Bury, Ixworth, Hadleigh, Bungay, Framlingham, &c.

XXXVIII. CORIANDRUM. Coriander.

1. C.* *sativum*, L.   Common C.   June.

About Ipswich and Framlingham. O.B.G.

ORD. XXXV.   ARALIACEÆ.—ARALIANTHS, THE ARALIA FAMILY.

I. HEDERA. Ivy.

1. H. *Helix*, L.   Common I.   October, Nov.

Abundant in woods and hedges, ruins, &c.

ORD. XXXVI. LORANTHACEÆ.—LORANTHANTHS, THE MISTLETOE FAMILY.

I. VISCUM. Mistletoe.

1. V. *album*, L.   Common M.   March—May.

Frequent on various trees.

ORD. XXXVII. CORNACEÆ.—CORNANTHS, THE CORNEL FAMILY.

I. CORNUS. Cornel.

1. C. *sanguinea*, L.   Common C.   (Dogwood.) June, July.

Hedges and thickets, abundant.

(Sect. II. Monopetalous.)

N.B. Bentham arranges this Section under Sub. Class 3. Corollifloræ.

ORD. XXXVIII.   CAPRIFOLIACEÆ.—LONICE-
RANTHS, THE HONEY-SUCKLE FAMILY.

I. ADOXA. Moscatel.

1. A. *Moschatellina*, L.   Tuberous M. (Gloryless.)
April, May.

Not unfrequent in woods and damp shady places; more
common on a heavy than a light soil.

II. SAMBUCUS. Elder.

1. S. *nigra*, L.   Common E.   June.

Abundant in woods, &c.   Lime-kilns, Eastgate, Bury.

2. S. *Ebulus*, L.   Dwarf E.   (Danewort.)   July,
August.

Langham, Brettenham, Boxford, Brampton, Rumburgh,
Hintlesham, and Sudbury.—Near Framlingham and
Parham on the way to Woodbridge.   Near Lowestoft,
Gorleston, and Halesworth. O.B.G.

III. VIBURNUM. Viburnum.

1. V. *Lantana*, L.   Mealy V.   (Wayfaring-Tree.)
May, June.

Woods and hedges about Bury, Hardwick and Hitcham.
Chedburgh. N.B.G.

2. V. *Opulus*, L.   Guelder-Rose V.   June, July.

Not uncommon in hedges and thickets, particularly near
water.   About Bury, Hitcham, Great Glemham and
Dennington.

IV. LONICERA. Honeysuckle.

1. L. *Periclymenum*, L. Common H. (Woodbine.)
June—September.

Plentiful in hedges and thickets.

2. L.* *Caprifolium*, L.   Perfoliate H.   May, June.

Not very general.   Hawstead Vale, Hitcham, Bungay, &c.

ORD. XXXIX. RUBIACEÆ.—RUBIANTHS.

N.B. British species are restricted to "The Stellate Tribe."

II. GALIUM. Galium.

1. G. *Cruciata*, Scop.   Crosswort G.   April—June.

    Not very general. Risby, Icklingham, Hitcham, Haughley, Great Glemham, Bungay, Whatfield, &c.

2. G. *verum*, L.   Yellow G.   (Ladies' Bedstraw.)
   June—September.

    Abundant on dry banks, walls, and waste places.

3. G. *palustre*, L.   Marsh G.   July, August.

    Common in marshes.

4. G. *uliginosum*, L.   Swamp G.   July, August.

    Not so general as the marsh G. but tolerably frequent in marshy places.

5. G. *saxatile*, L.   Heath G.   June—August.

    Plentiful in heathy ground.

6. G. *Mollugo*, L.   Hedge G.   July, August.

    Frequent in hedges, except near the sea, and in the North-west

7. G. *parisiense*, L.   Wall G.   June, July.

    Plentiful on the Vine-fields wall, Bury. Mildenhall and Icklingham Churchyard walls. Brandon. Little Barton Churchyard wall. O.B.G.   Wangford. Rev. J. Hemstead.

9. G. *Aparine*, L.   Cleavers G.   (Goosegrass.)
   June, July.

    Abundant in hedges, &c.

10. G. *tricorne*, With.   Corn G.   June—Oct.

    Local. Corn fields about Bury, Hardwick, Lavenham and Hitcham.   Badingham. J.W.G.

III. ASPERULA. Asperule.

1. A. *odorata*, L.   Woodruff A.   May, June.

    Damp woods at Hardwick, Hitcham, Chediston, &c.

2. A. *Cynanchica*, L. Small A. (Squinancy-wort.)
June, July.

Dry pastures at Bury, West Stow, Cavenham, Newmarket,
&c.

IV. SHERARDIA. Sherardia.

1. S. *arvensis*, L. Blue S. (Field-Madder.)
April—October.

Common in fields and waste places.

ORD. XL. VALERIANACEÆ.—VALERIANANTHS,
THE VALERIAN FAMILY.

I. CENTRANTHUS. Centranth.

1. C.* *ruber*, D. C. Red C. (Red-Valerian.)
June—September.

Rare. Walls of Rushbrook Hall.

II. VALERIANA. Valerian.

1. V. *dioica*, L. Marsh V. May, June.

Frequent in marshes.

2. V. *officinalis*, L. Common V. (All-heal.) June
—August.

Moist woods, sides of streams and ditches, plentiful.

III. VALERIANELLA. Cornsalad. (Fedia.)

1. V. *olitoria*, Poll. Common C. (Lambs'-Lettuce.)
April—June.

Common in cultivated ground.

4. V. *dentata*, Koch. Narrow-fruited C. June—
August.

Not very general. About Hitcham, Thorpe, Bungay and
Tuddenham.—Between Felixstow and Landguard Fort.
Phyt.—Near Bradwell. Hist. Y.—Near Halesworth. Br.
Fl.

ORD. XLI. DIPSACEÆ.—DIPSACANTHS, THE
TEASEL FAMILY.

I. DIPSACUS. Teasel.

1. D. *sylvestris*, L. Common T. August, Sept.

Common in damp hedges, river banks, &c.

2. D. *pilosus*, L. Small T. August, September.

Not unfrequent, except near the sea and in the North-west.

II. SCABIOSA. Scabious.

1. S. *succisa*, L. Blue S. (Devil's-bit.) July—October.

Abundant in moist pastures within the same limits as the former.

2. S. *Columbaria*, L. Small S. July, August.

Plentiful about Bury, Mildenhall, Hadleigh, and Bungay.

3. S. *arvensis*, L. Field S. (Knautia.) June—August.

Common on the borders of fields, &c.

ORD. XLII. COMPOSITÆ.—COMPOSITES OR AS-TERANTHS,—THE COMPOSITE FAMILY.

I. EUPATORIUM. Eupatory.

1. E. *cannabinum*, L. Common E. (Hemp-Agrimony. July—September.

Common on the banks of streams, &c.

II. TUSSILAGO. Coltsfoot.

1. T. *Farfara*, L. Common C. March, April.

Abundant in a clayey soil.

2. T. *Petasites*, L. Butterbur C. March—May.

Not unfrequent in low ground about streams.

III. ASTER. Aster.

1. A. *Tripolium*, L. Sea A. (Michaelmas-Daisy.) August, September.

Abundant in salt marshes.

IV. ERIGERON. Erigeron.

1. E. *acris*, L. Common E. (Fleabane.) July, August.

Frequent about Bury.—Livermere, Woolpit, Hitcham, Freston, Great Glemham, Mildenhall, and Tuddenham.

VI. SOLIDAGO. Goldenrod.

1. S. *Virga-aurea*, L. Common G. July—Sept.

Rather a local plant. Heathy ground about Lowestoft. Bungay, Dunwich and Snape.

VII. INULA. Inule.

1. I. *Helenium*, L. Elecampane I. July, Aug.

Rare. Stoven, by the Cherry Tree. Mr. Wigg.—Ufford, Rattlesden, Parham, Mettingham, Sibton, Bramfield, and Heveningham. O.B.G.

2. I. *crithmoides*, L. Samphire I. (Golden-Samphire.) July, August.

" On the sea coast." O.B.G.

3. I. *Conyza*, D. C. Rigid I. (Ploughman's Spikenard.) August—October.

Sparingly in a clayey soil. About Hawstead, Little Welnetham, Hitcham, Bungay, Great Glemham, Pakenham, and Cavenham.

4. I. *dysenterica*, L. Common I. (Pulicaria.) July—September.

Common in wet places.

5. I. *Pulicaria*, L. Small I. (Pulicaria vulgaris.) August, September.

Framlingham, very scarce. Hist. F.

VIII. BELLIS. Daisy.

1. B. *perennis*, L. Common D. Feb.—Oct.

Abundant everywhere.

IX. CHRYSANTHEMUM. Chrysanthemum.

1. C. *Leucanthemum*, L. Ox-eye C. June, July.

Common in pastures, railway embankments, &c.

2. C. *segetum*, L. Corn C. (Corn-Marigold.) June—October.

Plentiful about Tuddenham, Higham, Barrow, Pakenham, Wadringfield, Sudbourne, and Ipswich.

3. C. *Parthenium*, Pers. Feverfew C. (Pyrethrum.) July—September.

Apparently wild about Bungay.—Hitcham, Great Glemham, Bury, &c.

4. C. *inodorum,* S. Scentless C. (Scentless Mayweed.) June—November.

Very common everywhere.

X. MATRICARIA. Matricary.

1. M. *Chamomilla,* L. Common M. (Wild-Chamomile.) June—August.

Probably a common plant. Recorded from Hitcham, Bungay, Bergholt, and Ipswich.

XI. ANTHEMIS. Chamomile.

1. A, *Cotula,* L. Fetid C, (Stinking-Mayweed.) June—September.

Plentiful in fields, waste places, &c.

2. A. *arvensis,* L. Corn C. June—August.

Not common. Fields about Bury, Hardwick, and Hitcham. —Ipswich. W.L.N.—Framlingham. Hist. F.

3. A. *nobilis,* L. Common C. July—September.
Rather scarce. About Bungay.— Blundeston, Bradwell, and other Commons about Yarmouth. O.B.G.—Near Lowestoft. Eng. Fl.

XII. ACHILLÆA. Achillæa.

1. A. *Ptarmica,* L. Sneezewort A. July, Aug.

Tuddenham heath. Pakenham, Bungay, and Great Glemham.—Framlingham. Hist. F.

2. A. *Millefolium,* L. Milfoil A. (Yarrow.) June—September.

Abundant everywhere.

XIII. DIOTIS. Diotis

1. D. *maritima,* Cass. Sea D. (Santolina.) Aug., September.

Very rare. Near Landguard Fort.—Aldboro, Orford, and Dunwich. O.B.G. Occasionally at Pakefield and Southwold.

XIV. TANACETUM. Tansy.

1. T. *vulgare,* L. Common T. August.

Not unfrequent in waste places and roadsides. About Bury,
Lowestoft, Great Glemham, Bungay, &c.

XV. ARTEMISIA. Artemisia.

1. A. *campestris*, L. Field A. August, Sept.

Rare. Mildenhall, Brandon, and Wangford. Near the
Flint Works, Icklingham. About Barton Mills, Elden,
and Icklingham heath. O.B.G.

2. A. *maritima*, L. Sea A. August, September.

Frequent in salt marshes and sea wastes. Lowestoft,
Aldborough, Walton ferry, &c.

3. A. *vulgaris*, L. Common A. (Mugwort.)
July—September.

Common in hedges and waste places.

4. A. *Absinthium*, L. Wormwood A. August,
September.

Chalk pit, near Hospital, Bury. Bungay, Benhall, Bentley,
and Somersham.—Dunwich. J. W. G.

XVI. GNAPHALIUM. Cudweed.

1. G. *dioicum*, L. Mountain C. (Cat's-Ear.)
(Antennaria.) June, July.

Rare. Newmarket heath, plentiful. Culford and Cavenham heaths.

2. G. *margaritaceum*, L. Pearl C. (Antennaria.)
August.

Very rare. Gravel banks at Bentley.

3. G. * *luteo-album*, L. Jersey C. July, August.

Very rare. Eriswell.

4. G. *sylvaticum*, L. Wood C. (G. rectum.)
July—September.

Roadside from Farnham to Snape; Belton common and
Fritton ; hedge by Corton wood ; Road between Yarmouth
and Beccles. O.B.G.

6. G. *uliginosum*, L. Marsh C. July—Sept.

Frequent in muddy waste places, especially on a clayey
soil. Hitcham, Great Glemham, Naughton, Bungay,
and Mildenhall.

7. G. *germanicum*, Willd. Common C. (Filago.)
    July—September.

    Abundant in waste and cultivated ground.

8. G. *arvense*, Willd. Field C. (Filago minima.)
    June—September.

    Plentiful at Mildenhall and Tuddenham.—Great Glemham
    and Bungay.—About Westleton. J.W.G.

9. G. *gallicum*, Huds. Narrow G. (Filago.)
    July—September.

    "Suffolk." Br. Fl.

    ## XVII. SENECIO. Senecio.

1. S. *vulgaris*, L. Groundsel S. (Simpson.) Jan.
    —December.

    Abundant in cultivated ground.

2. S. *viscosus*, L. Viscous. S. July, August.

    Aldborough and Lowestoft. Old brick grounds, Newmarket
    road, Bury, occasionally.

3. S. *sylvaticus*, L. Wood S. July—September.

    Frequent about Tuddenham and Icklingham.—Bungay,
    Snape, and Sizewell.

4. S.* *squalidus*, L. Squalid S. June—October.

    Plentiful on old walls at Bury.

5. S. *aquaticus*, Huds. Water S. July, August.

    Common in marshes and wet places. Possibly a var. of
    the next.

6. S. *Jacobæa*, L. Ragwort S. July—September.

    Abundant everywhere.

7. S. *erucæfolius*, L. Narrow-leaved S. (S. ten-
    uifolius.) July, August.

    Frequent in most districts.

8. S. *paludosus*, L. Fen S. June, July.

    Very rare. Lakenheath Fen by Wangford.

10. S. *palustris*, D. C. Marsh S. (Cineraria.)
    June, July.

    Very rare. Brandon. Geldeston Fen.—Worlingham Com-
    mon and Haddiscoe. O.B.G.—Belton. N.B.G.

11. S. *campestris*, D.C. Field S. (Cineraria integrifolia.) May, June.

Devil's-ditch, Newmarket.

XIX. BIDENS. Bidens.

1. B. *cernua*, L. Nodding B. (Bur-Marigold.) July—October.

Not uncommon in watery places. Livermere, Bungay, and Farnham.—Framlingham. Hist. F.

2. B. *tripartita*, L. Three-cleft B. July—Sept.

Not so frequent as the former. Bury, Fornham, Hitcham, and Benhall Green.—Framlingham. Hist. F.

XXI. ARCTIUM. Arctium.

1. A. *Lappa*, L. Common Burdock. July, Aug.

Roadsides and waste places, abundant.

XXII. SERRATULA. Sawwort.

1. S. *tinctoria*, S. Common S. August.

Reported to grow in the county.

XXIV. CARDUUS. Thistle.

1. C. *Marianus*, L. Milk T. (Silybum.) July.

Not unfrequent about Bury, Gedding, Stowlangtoft, Aldborough and Ipswich.

2. C. *nutans*, L. Musk T. May—October.

Common on roadsides and waste ground.

3. C. *acanthoides*, L. Welted T. June—August.

Not very general. About Bury, Hitcham, Great Glemham, Felixstow, Bungay, &c.

4. C. *pycnocephalus*, Jacq. Slender T. (C. tenuiflorus.) June—August.

Waste ground near the sea, frequent.

5. C. *lanceolatus*, L. Spear T. (Cnicus.) July, August.

Abundant on road sides and waste places.

6. C. *palustris*, L. Marsh T. (Cnicus.) July.

Low meadows and marshes, common.

7. *C. arvensis*, Curt. Creeping T. (Cnicus.) July.
Abundant in waste and cultivated ground.

8. *C. eriophorus*, L. Woolly T. (Cnicus.) July, August.
Not general. About Hitcham, Ottley, and Eye.—Near Framlingham, Clare, and Debenham. O.B.G.

11. *C. pratensis*, Huds. Meadow T. (Cnicus.) June—August.
Boggy ground on Tuddenham heath, and very plentiful in deep bogs near the river.—Wangford.—Marsh at Bradwell. O.B.G.—Belton, rare. Hist. Y.

12. *C. acaulis*, L. Dwarf T. (Cnicus.) July—September.
Not unfrequent in pastures. About Bury, Hitcham, Little Glemham Park, &c.

XXV. ONOPORDON. Onopord.

1. O. *Acanthium*, L. Common O. (Scotch or Cotton Thistle.) August.
Not uncommon about Bury, Sudbury, Ipswich, Sizewell, and Mildenhall.

XXVI. CARLINA. Carline.

1. C. *vulgaris*, L. Common C. June—October.
Frequent. Nowton, Hitcham, Bungay, Framlingham, Higham, Tuddenham, &c.

XXVII. CENTAUREA. Centaurea.

1. C. *nigra*, L. Black C. (Knapweed.) June—September.
Abundant in pastures, &c.

2. C. *Scabiosa*, L. Greater C. July—Sept.
Common on hedge-banks, waste places, &c.

3. C. *Cyanus*, L. Corn C. (Blue-Bottle. Cornflower.) June—August.
Frequent in corn fields, in a light sandy soil. Higham, Tuddenham, &c.—Hitcham and Aldborough.

5. C. *Calcitrapa*, L. Star-Thistle C. July—Oct.
Near the fish-houses North Denes Lowestoft, plentiful. About Orford.—Common about Yarmouth. Hist. Y.

6. C. *solstitialis*, L. Yellow C. July—Sept.

Occasionally. Waste ground, Lowestoft Docks. Borders of fields at Bury, Rougham, Welnetham, Gunton, &c.

XXVIII. TRAGOPOGON. Salsify.

1. T. *pratense*, L. Meadow S. (Yellow Goat's-Beard.) June, July.

Not unfrequent, especially in a heavy soil.

2. T.* *porrifolium*, L. Purple S. May, June.

Meadows between Hawstead and Great Welnetham.

XXIX. HELMINTHIA. Helminth.

1. H. *echioides*, Gærtn. Ox-tongue H. June—October.

Not very general. Hedge-banks, borders of fields, &c., about Bury, Pakenham, Chevington, Hitcham, Great Glemham, Bungay, &c.

XXX. PICRIS. Picris.

1. P. *hieracioides*, L. Hawk-weed P. June—Oct.

Not unfrequent about Hardwick, Felsham, Hitcham, Lavenham, Great Glemham, &c.

XXXI. LEONTODON. Hawkbit.

1. L. *hispidus*, L. Common H. (Apargia.) June—September.

Common in all districts.

2. L. *autumnalis*, L. Autumnal H. (Apargia.) August.

Abundant everywhere.

3. L. *hirtus*, L. Lesser H. (Thrincia.) July, Aug.

Common in a gravelly soil, dry pastures, &c.

XXXII. HYPOCHŒRIS. Hypochœre.

1. H. *glabra*, L. Glabrous H. June—October.

Not very frequent. Mildenhall, Tuddenham, Brandon, Barham, Bungay, Aldborough, and Benhall Green. —Great Barton. O.B.G.

2. H. *radicata*, L. Long-rooted H. (Cat's-Ear.) July, August.

Abundant everywhere.

3. H. *maculata*, L. Spotted H. July, August.

Rare. Risby heath, on a chalky bank near the fir plantation. O.B.G.—Newmarket and Icklingham. N.B.G.

XXXIII. LACTUCA. Lettuce.

1. L. *muralis*, Fresen. Wall L. (Prenanthes.) June—August.

About Chevington and Groton.—Framlingham. Hist. F.

2. L. *scariola*, L. Prickly L. April—August.

Newmarket. N.B.G. *Var. virosa*, Ixworth Thorp, Rattlesden, Great Glemham, Bungay, Saxmundham, Sproughton, Hadleigh, &c.

3. L. *saligna*, L. Willow L. July, August.

Rare. Salt marshes at Aldborough.

XXXIV. SONCHUS. Sow-thistle.

1. S. *arvensis*, L. Corn S. August, September.

A common weed in corn-fields.

2. S. *palustris*, L. Marsh S. July, August.

Very rare. Sides of ditches dividing arable from marsh land north of Oulton Dyke.—Marshes near Burgh Staithe.—River side at Beccles and Worlingham, and ditches under Burgh Castle. O.B.G.

3. S. *oleraceus*, L. Common S. June—August.

Both varieties are common in cultivated ground.

XXXV. TARAXACUM. Dandelion.

1. T. *Dens-Leonis*, Desf. Common D. March—October.

Abundant everywhere.

XXXVI. CREPIS. Crepis.

1. C. *taraxacifolia*, Thuil. Beaked C. (Borckhausia.) June, July.

Rare. Great Glemham, Framlingham, and Marlesford.

2. C. *fœtida*, L. Fetid C. (Borckhausia.) June, July.

Very rare. About Claydon, Coddenham, and Saxham. O.B.G.

3. C. *virens*, L. Smooth C. (C. tectorum.) June —September.

Common everywhere.

4. C. *biennis*, L. Rough C. June, July.

About Framlingham. Sparingly in fields about Bury. —Parham. O.B.G.

XXXVII. HIERACIUM. Hawkweed.

1. H. *Pilosella*, L. Mouse-ear H. May—Aug.

Abundant on dry banks.

3. H. *murorum*, L. Wall H. June—August.

Near Framlingham. Hist. F.

5. H. *umbellatum*, L. Umbellate H. Aug, Sept.

Stowlangtoft.—Dunwich Cliff.

XXXVIII. CICHORIUM. Chicory.

1. C. *Intybus*, L. Wild C. (Succory.) July— October.

Not uncommon on roadsides and waste ground.

XXXIX. ARNOSERIS. Arnoseris.

1. A. *pusilla*, Gærtn. Dwarf A. (Lapsana.) June, July.

Rare. Lakenheath. Broom Close, Barham.—Framling-ham. Hist. F.—Mildenhall. J.T.—Fornham, Timworth, Harleston, Dingle, Westleton, and near the Alms houses, Woodbridge. O.B.G

XL. LAPSANA. Lapsane.

1. L. *communis*, L. Common L. (Nipplewort.) July—September.

Common everywhere.

ORD. XLIII. CAMPANULACEÆ.—CAMPANU-
LANTHS, THE CAMPANULA FAMILY.

II. JASIONE. Jasione.

1. **J. montana, L.** Sheep's-bit J. June—September.

Not unfrequent in heathy ground. Icklingham, Farnham,
Great Glemham, Aldborough, Framlingham, &c.

IV. CAMPANULA. Campanula.

1. **C. glomerata, L.** Clustered C. July, August.

Not very common. Dry pastures and thickets about
Bury, Westley, Saxham, Barrow, Polstead, &c.

2. **C. Trachelium, L.** Nettle-leaved C. July—
September.

Frequent in woods and hedges, except near the sea, and in
the North-west. Hardwick, Hawstead, Great Glemham,
Hitcham, Drinkstone, Rattlesden, Boxford, &c.

3. **C. latifolia, L.** Giant C. July, August.

Recorded from Shipmeadow, Linstead Parva and elsewhere near Halesworth in O.B.G. Also from Chediston
in 1846.

5. **C.* Rapunculus, L.** Rampion C. July, August.

Hedges to the right of the road a mile from Beccles
towards Halesworth.

7. **C. rotundifolia, L.** Harebell C. July, Sept.

Frequent in a dry heathy soil.

9. **C. hybrida, L.** Corn C. June—September.

Plentiful in corn-fields about Bury, Hitcham, Benhall, &c.

ORD. XLIV. ERICACEÆ.—ERICANTHS, THE
HEATH FAMILY.

I. VACCINIUM. Vaccinium.

4. **V. Oxycoccos, L.** Cranberry V. June.

Very rare. Wangford.—Worlingham Common. O.B.G.

VII. ERICA. Heath.

1. E. *vulgaris*, L. Common H. (Ling.) (Calluna.) June—August.

Abundant on heaths, sometimes with white flowers.

2. E. *cinerea*, L. Scotch H. July—September.

With the preceding and equally common.

3. E. *Tetralix*, L. Cross-leaved H. July, Aug.

Not unfrequent on heaths, but not usually plentiful. Lowestoft, Lound, Dunwich, Ipswich, Tuddenham, Wangford, &c.

VIII. PYROLA. Wintergreen.

2. P. *rotundifolia*, L. Larger W. July· –Sept.

Very scarce. Ashen Spring, near the round house, Theberton.—Bradwell Common, among Furze. O.B.G.— Wood at Middleton. Eng. Fl.

IX. MONOTROPA. Monotrope.

1. M. *Hypopitys*, L. Common M. (Yellow-Bird's-nest.) June, July.

Rare. Plentiful in the arboretum of the Marquis of Bristol, Ickworth Park.—Near Bungay.

SUB-CLASS III. COROLLIFLORÆ.—COROLLI-FLORES.

ORD. XLV. PRIMULACEÆ.—PRIMULANTHS, THE PRIMROSE FAMILY.

I. HOTTONIA. Hottonia.

1. H. *palustris*, L. Water H. (Water-Violet. Featherfoil.) May, June.

Not unfrequent in ditches in almost all parts of the county.

II. PRIMULA. Primrose.

1. P. *veris*, L. Common P. April, May.

a. *vulgaris*, Primrose. Common in hedgebanks and woods —varying occasionally with pink flowers.

b. *veris*, Cowslip. Common in pastures and hedges.

c. *elatior*, Oxlip. Plentiful in moist pastures and hedge-banks at Hawstead, Hitcham, Finborough, &c.

2. P. *farinosa*, L.   Mealy P.   June, July.

Very rare.   Cavenham Severals, and Bergholt.

IV. LYSIMACHIA.   Lysimachia.

1. L. *vulgaris*, L.   Common L.   (Loosestrife.)
July, August,

Not a very common plant.   Ditch banks and thickets in marshes about Oulton, Somerleyton, Bungay and Bergholt.—Near Icklingham. Hist. Bury.

3. L. *nummularia*, L.   Moneywort L.   June, July.

Frequent in damp woods and sides of ditches at Bury, Hitcham, Great Glemham, Bungay, Rattlesden, &c.

4. L. *nemorum*, L.   Wood L.   May—Aug.

Not uncommon in woods, hedgebanks, &c., about Lowestoft, Fritton, Freston, Great Glemham, Hitcham, Naughton, Bungay, Haughley and Rushbrooke.

VI. GLAUX.   Glaux.

1. G. *maritima*, L.   Sea G.   (Sea-Milkwort.)
June, July.

Salt marshes and sea wastes, plentiful.

VII. ANAGALLIS.   Pimpernel.

1. A. *arvensis*, L.   Common P.   (Shepherd's-weatherglass.)   May—October.

The red variety common everywhere ; the blue occurs at Little Welnetham.

2. A. *tenella*, L.   Bog P.   July, August.

Abundant in boggy ground about Lowestoft, Snape, Sizewell, Benhall, Framlingham, Mildenhall, Tuddenham, &c.

VIII. CENTUNCULUS.   Centuncule.

1. C. *minimus*, L.   Small C.   (Chaffweed.)   June, July.

Rare.   About Lowestoft and Wangford.   Belton and Herringfleet Commons.—About Oulton Broad. Hist. Y.

IX. SAMOLUS.   Samole.

1. S. *Valerandi*, L.   Brookweed S.   June—Sept.

Not unfrequent in wet places about Lowestoft, Snape, Great Glemham, Benhall, Ipswich, Framlingham, Bungay, Tuddenham, &c.

ORD. XLVI. LENTIBULACEÆ.—PINGUICU-
LANTHS, THE PINGUICULA FAMILY.

I. PINGUICULA. Butterwort.

1. P. *vulgaris*, L. Common B. May—July.

Boggy ground about Stowlangtoft, Mildenhall, Tuddenham, &c.

II. UTRICULARIA. Bladderwort.

1. U. *vulgaris*, L. Common B. June, July.

Frequent in boggy pools and ditches. Tuddenham, near
the river. Mildenhall and Hopton. Ditches in Pakenham Fen.

2. U. *minor*, L. Lesser B. June—September.

In the same situations as the preceding and often in com-
pany with it. Herringfleet, St Olave's Bridge, Lackford,
Tuddenham, Cavenham, Lakenheath and Brandon.—
Lound and Gorleston. O.B.G.—Belton. N.B.G.

Var. intermedia, Burgh Common. O.B.G.

ORD. XLVII. AQUIFOLIACEÆ.—ILICANTHS,
THE HOLLY FAMILY.

I. ILEX. Holly.

1. I. *Aquifolium*, L. Common H. May, June.

Frequent in woods and hedges.

ORD. XLVIII. JASMINACEÆ.—JASMINANTHS,
THE JESSAMINE FAMILY.

I. FRAXINUS. Ash.

1. F. *excelsior*, L. Common A. April, May.

Common in woods.

II. LIGUSTRUM. Privet.

1. L. *vulgare*, L. Common P. June, July.

Hedges and thickets, not unfrequent.

ORD. XLIX. APOCYNACEÆ.—APOCYNANTHS,
THE PERIWINKLE FAMILY.

I. VINCA. Periwinkle.

1. **V.*** *major*, L. Larger P. April, May.

Claydon, by the roadside. About Bungay.—Near Hawstead Green. O.B.G.

2. **V.** *minor*, L. Lesser P. April, June.

Frequent on hedgebanks and in woods. Hunston, Rushbrooke, the Bradfields, Welnetham, Bergholt, Naughton,
&c.

ORD. L. GENTIANACEÆ.—GENTIANANTHS,
THE GENTIAN FAMILY.

II. ERYTHRÆA. Erythræa.

1. **E.** *Centaurium*, Pers. Common E. (Centaury.)
June—October.

a. Common in dry pastures, &c.
c. Near the sea. Lowestoft Denes. Near Gorleston pier.
Aldborough, &c.

III. GENTIANA. Gentian.

1. **G.** *Pneumonanthe*, L. Marsh G. Aug., Sept.

Very rare. Carlton Colville; Hopton and Corton heaths.
O.B.G.

4. **G.** *Amarella*, L. Autumn G. July—Sept.

Not very common. Thurston heath; Rougham and Bradfield.—Little Saxham and Barrow Bottom. O.B.G.

5. **G.** *campestris*, L. Field G. Aug.—October.

Rare. Rougham and Hartest. Bank on Icklingham heath.
—Cavenham Severals. O.B.G.

IV. CHLORA. Chlora.

1. **C.** *perfoliata*, L. Perfoliate C. (Yellowwort.)
June—September.

Frequent in a chalky or clayey soil. Hardwick, Cheving-
ton, Saxham, Haughley, Norton, Elmswell, Rougham,
Felsham, Brettenham, Hitcham, Lavenham, Naughton,
Great Glemham, Rendham, Marlesford, Badingham,
Bungay.

V. MENYANTHES. Buckbean.

1. M. *trifoliata*, L. Common B. (Bogbean. Marsh Trefoil.) May—July.

Boggy ground at West Stow, Icklingham, Tuddenham, Higham, Benhall Green, Bungay, &c.

ORD. LII.   CONVOLVULACEÆ.—CONVOLVU-LANTHS. THE CONVOLVULUS FAMILY.

I. CONVOLVULUS. Convolvulus.

1. *arvensis*, L. Lesser C. (Bindweed.) June, July.

Very common in fields, hedgebanks, &c.

2. C. *sepium*, L. Larger C. (Calystegia.) June —August.

Common in damp hedges and thickets.

3. C. *Soldanella*, L. Sea C. (Calystegia.) June —August.

Not unfrequent on the sandy coast.

II. CUSCUTA. Dodder.

1. C. *europæa*, L. Greater D. July—Sept.

Not very general, but abundant on Hops in hedges a short distance beyond the Gaol, Bury. Whatfield and Semer. —Worlingham heath. O.B.G.

2. C.* *Epilinum*, Weihe. Flax D. August.

On Flax at Framlingham.

3. C. *Epithymum*, L. Lesser D. August.

Frequent and plentiful on Thyme, Heath, Clover, &c.

ORD. LIII.   BORAGINACEÆ.—BORAGINANTHS, THE BORAGE FAMILY.

I. ECHIUM. Echium.

1. E. *vulgare*, L. Common E. (Viper's-Bugloss.) June, July.

Common, especially in a light soil, in fields, waste places, &c.

IV. LITHOSPERMUM. Lithosperm.

1. **L.** *arvense,* **L.**   Corn L.   (Corn-Gromwell.) (Bastard-Alkanet.)   May, June.

Less frequent than the next species. Fields about Bury, Hitcham, Bungay, Framlingham, &c.

2. **L.** *officinale,* **L.**   Common L.   June.

Hedges :—tolerably common about Hardwick, Hitcham, Great Glemham, &c.

3. **L.** *purpureo-cæruleum,* **L.**   Creeping L.   June, July.

Very rare. Bergholt.

V. MYOSOTIS. Myosote.

1. **M.** *palustris,* With.   Water M.   (Forget-me-not.)   June—August.

Abundant in watery places.   *Var. cæspitosa,*—Bury, Hitcham, Snape, Bungay, Tuddenham, &c.

2. **M.** *sylvatica,* Hoffm.   Wood M.   May—Aug.

Not common. Woods at Hitcham, Ottley, and Bricet.

3. **M.** *arvensis,* Roth.   Field M.   June—August.

Plentiful on hedgebanks, in fields, &c.

4. **M.** *collina,* Hoffm.   Early M.   April, May.

Frequent on dry banks in most districts.

5. **M.** *versicolor,* Pers.   Changing M.   (M. scorpioides.)   April—June.

Less common than the preceding. Haberden, Bury; Rougham, Great Glemham, Bungay, West Stow, and Tuddenham.

VI. ANCHUSA. Alkanet.

1. **A.*** *officinalis,* **L.**   Common A.   June, July.

Rare. Bergholt.

2. **A.*** *sempervirens,* **L.**   Green A.   May, June.

Great Livermere, Bergholt, &c. Walpole, Eng. Fl.

VII. LYCOPSIS. Bugloss.

1. **L.** *arvensis,* **L.**   Small B.   June, July.

Common on banks and in cultivated ground.

VIII. SYMPHYTUM. Comfrey.

1. S. *officinale*, L.   Common C.   May, June.

River bank at Santon Downham. About Mildenhall, Bergholt, Bungay, and Horringer.—Framlingham. Hist. F.
—Badingham and Laxfield. J.W.G.

IX. BORAGO. Borage.

1. B.* *officinalis*, L.   Common B.   June, July.

Waste ground in various places.

X. ASPERUGO. Asperugo.

1. A. *procumbens*, L.   German A.   (Madwort.)
June, July.

Very rare.   Wangford near Brandon.—Near Newmarket.
O.B.G.

XI. CYNOGLOSSUM. Hound's-Tongue.

1. C. *officinale*, L.   Common H.   June, July.

Roadsides and waste ground about Bury, Barton, Hitcham,
Great Glemham, Lowestoft, West Stow, &c.

ORD. LIV. SOLANACEÆ.—SOLANANTHS, THE SOLANUM FAMILY.

I. DATURA. Datura.

1. D.* *Stramonium*, L.   Thorn-Apple D.   July—
October.

Waste ground at Santon Downham.—On Fritton heath and
adjoining hedges, copiously. O.B.G.

II. HYOSCYAMUS. Henbane.

1. H.* *niger*, L.   Common H.   June—August.

Not unfrequent in hedges, waste ground, and roadsides
near towns and villages.   Bury, Barton, Cockfield,
Hitcham, Lowestoft Denes, Bungay, Aldborough, Snape,
Felixstow, Icklingham, &c.

III. SOLANUM. Solanum.

1. S. *Dulcamara*, L.   Bittersweet S.   (Woody-
Nightshade.)   June—August.

Common in moist hedges and thickets.

2. S. *nigrum*, L. Black S. June—November.
Abundant in cultivated and waste ground.

IV. ATROPA. Atropa.

1. A. *Belladonna*, L. Deadly A. (Dwale. Deadly-Nightshade.)
Ruins at Bury and Framlingham. Lane between Risby and Flempton. Near Mildenhall.

ORD. LV. OROBANCHACEÆ.—OROBANCHANTHS, THE BROOMRAPE FAMILY.

I. OROBANCHE. Broomrape.

1. O. *major*, L. Great B. May—July.
Occasionally on Broom and Furze. About Hadleigh, Bungay, &c.
2. O. *caryophyllacea*, Sm. Clove-scented B. July.
Very rare. Semer.
4. O. *elatior*, Sutt. Tall B. June—August.
Rare. Sutton.—About Bury. O.B.G.
5. O. *minor*, L. Lesser B. June—August.
Plentiful in most districts on clover, &c.

ORD. LVI. SCROPHULARINEÆ.—SCROPHULA-RIANTHS, THE SCROPHULARIA FAMILY.

I. VERBASCUM. Mullein.

1. V. *Thapsus*, L. Great M. June—August.
Frequent in hedges, etc., about Bury, Rougham, Hitcham, Great Glemham, Bungay, etc.
2. V. *virgatum*, With. Twiggy M. August.
Scarce. About Semer and Ixworth.
4. V. *nigrum*, L. Dark M. June—October.
Common in hedges, roadsides, etc. About Bury, Hitcham, Hadleigh, Bungay, Westleton, and Ipswich.

5. V. *Lychnitis*, L. White M. July, August.

Rare. Roadside between Ampton and Livermere.—About Framlingham. Hist. F.

6. V. *pulverulentum*, Vill. Hoary M. (V. flocco-sum.) July.

Not unfrequent in chalk pits, and roadsides, near Bury.—Dunwich.—At the back of Browston Hall; near Mutford bridge; Herringfleet Common. O.B.G.—Fritton wood. Hist. Y.

II. ANTIRRHINUM. Snapdragon.

1. A.* *majus*, L. Great S. July—Sept.

Old walls at Bury, Fornham, Ipswich, and Bergholt.

2. A. *Orontium*, L. Lesser S. July—Oct.

Cultivated fields near Bury, Great Glemham, Bergholt, and Bungay.—Theberton. J.W.G.—Belton. Hist. Y.—Framlingham. Hist. F.

III. LINARIA. Linaria.

1. L. *vulgaris*, Mœnch. Common L. (Toad-flax.) July—October.

Abundant in hedges, borders of fields, etc.

Var. Peloria, Frequent about Bury. A very pretty variety occurs in the chalk pit, near Hospital, Bury, with pale cream-colored flowers, pencilled with purple, and with yellow palate.

5. L. *minor*, Desf. Lesser L. May—October.

Frequent in cultivated ground about Hardwick, Whepstead, Hitcham, Bardwell, Newmarket and Lakenheath.—Aldborough and Iken. O.B.G.

6. L.* *Cymbalaria*, Mill. Ivy L. May—Sept.

Plentiful on old walls and ruins at Bury, Bungay, Framlingham, etc.

7. L. *spuria*, Mill. Round-leaved L. July—Nov.

Frequent in damp fields, etc., about Hardwick, Chedburgh, Little Welnetham, Hitcham, Great Glemham, Bungay, Halesworth and Lakenheath.—Badingham. J.W.G.—Framlingham. O.B.G.

8. L. *Elatine*, Desf. Pointed L. (Sharp-pointed-Fluellin.) July—November.

Equally distributed with the preceding, and frequently growing with it.

G

IV. SCROPHULARIA. Scrophularia.

1. S. *nodosa*, L. Knotted S. (Figwort.) June—August.

Moist ground and shady places, frequent.

2. S. *aquatica*, L. Water S. June—September.

Abundant by the sides of streams and ditches.

VI. LIMOSELLA. Limosel.

1. L. *aquatica*, L. Common L. July—Sept.

Rare. Moist places on Lowestoft Denes, near the fish-houses. About Mildenhall.

VIII. DIGITALIS. Foxglove.

1. D.* *purpurea*, L. Purple F. May—August.

About Aldborough, Bentley and Bungay.

IX. VERONICA. Veronica. (Speedwell.)

1. V. *spicata*, L. Spiked V. July, August.

Tuddenham and Cavenham heaths. Brandon.—Culford and Risby heaths. O.B.G.

4. V. *serpyllifolia*, L. Thyme-leaved V. May, June.

Plentiful in pastures, hedge-banks, etc.

5. V. *officinalis*, L. Common V. May—July.

Frequent in a dry, heathy soil.

6. V. *Anagallis*, L. Water V. July, August.

Common in ditches, etc.

7. V. *Beccabunga*, L. Brooklime V. May—Sept.

Wet places and sides of ditches, common.

8. V. *scutellata*, L. Marsh V. July, August.

Not very general or plentiful. Boggy ground at West Stow, Tuddenham, Mildenhall, Whatfield, Bungay, Ottley and Framlingham.

9. V. *montana*, L. Mountain V. April—July.

Link woods, Rushbrook.—Woods at Chediston, Great Glemham and Bungay.

10. V. *Chamœdrys*, L. Germander V. May, June.
Plentiful everywhere.

11. V. *hederifolia*, L. Ivy V. March—August.
Abundant everywhere.

12. V. *agrestis*, L. Procumbent V. April—Sept.
Common in cultivated ground.
Var. polita,—About Bury, Rougham, Stowlangtoft, Great Glemham, Bungay, etc.

13. V.* *Buxbaumii*, Ten. Buxbaum's V. April—September.
Frequent in gardens and cultivated fields about Bury, Hitcham, Stowlangtoft, Woolpit, etc.

14. V. *arvensis*, L. Wall V. April—July.
Common on wall tops, dry pastures, etc.

15. V. *verna*, L. Vernal V. April, May.
Very rare. Mildenhall, Brandon, Lakenheath, Thetford, Icklingham and West Stow.—In the Rye at Wordwell. O.B.G.

16. V. *triphyllos*, L. Fingered V. April.
Rare. West Stow, Icklingham, Lakenheath and Thetford.—Livermere, Aldborough, and Barham heath. O.B.G.

X. BARTSIA. Bartsia.

1. B. *Odontites*, Huds. Red B. June—August.
Common in fields and waste places.

XI. EUPHRASIA. Eyebright.

1. E. *officinalis*, L. Common E. May—Sept.
Moist pastures, plentiful.

XII. RHINANTHUS. Rattle.

1. R. *Crista-galli*, L. Common R. (Yellow Rattle.) May—July.
Abundant in pastures, marshes, etc.

XIII. PEDICULARIS. Pedicularis.

1. P. *palustris*, L. Marsh P. (Red-Rattle.) May—September.
Very general in marshy ground.

2. P. *sylvatica*, L.   Common P.   (Lousewort.)
April—July.

Not unfrequent in moist peaty ground.

XIV. MELAMPYRUM. Melampyre.

1. M. *cristatum*, L.   Crested M.   July.

Not uncommon in underwoods and thickets about Hardwick,
Whepstead, Westley and Chedburgh.

2. M.* *arvense*, L.   Purple M.   (Cowwheat.)
June—August.

Occasionally at Hitcham.

3. M. *pratense*, L.   Common M.   May—August.

Common in woods and thickets.

ORD. LVII.—LABIATÆ.—LABIATES OR LAMIANTHS,
THE LABIATE FAMILY.

I. SALVIA.   Sage.   (Clary.)

1. S.* *pratensis*, L.   Meadow S.   June, July.

Stowlangtoft.

2. S. *verbenaca*, L.   Wild S.   May—August.

Waste places, hedgebanks, etc., about Bury, Hitcham,
Bungay, Framlingham, and Marlesford.—Bawdsey Ferry
and Orford Castle.

II. LYCOPUS.   Lycopus.

1. L. *europæus*, L.   Common L.   (Gipsy-wort.)
June—September.

Common by the sides of ditches, etc.

III. MENTHA.   Mint.

1. M. *sylvestris*, L.   Horse M.   August, Sept.

Not very general. About Tuddenham, Brantham, Hitcham,
Bungay, Stowmarket and Westhall.—Aldborough, Burgh
Castle, Gillingham and Browston. O.B.G.

2. M. *rotundifolia*, L.   Round-leaved M.   August,
September.

Very rare. Chediston.

4. M.* *piperita*, Sm. Pepper M. August, Sept.
Waste ground, occasionally, in various places.

5. M. *aquatica*, L. Water M. August, Sept.
Abundant in wet ditches and marshy ground.

6. M. *sativa*, L. Whorled M. July, August.
Plentiful in Pakenham Fen, and, probably a common plant elsewhere.

7. M. *arvensis*, L. Corn M. August, September.
Very common in fields and waste ground.

8. M. *Pulegium*, L. Pennyroyal M. August, September.
Rare. About Bungay.—Parham, Benhall, Yoxford, Middleton and Belton. O.B.G.

IV. THYMUS. Thyme.

1. T. *Serpyllum*, L. Wild T. June—August.
Frequent on dry banks, hilly pastures, etc.

V. ORIGANUM. Marjorum.

1. O. *vulgare*, L. Wild M. July—September.
Rare. On a dry bank at Hitcham.

VI. CALAMINTHA. Calamint.

1. C. *Acinos*, Clairv. Field C. (Basil-Thyme.) July.
Frequent about Risby, Tuddenham, Mildenhall, Thurston, Orford, Aldborough and Freston.

2. C. *officinalis*, Mænch. Common C. July—September.
a. *Thymus Nepeta*, Near Kennett Bell. O.B.G.
b. *Thymus Calamintha*, Not uncommon about Bury, Semer, Bungay, Framsden, Brandeston and Little Glemham.—Gorleston. Hist. Y.—Peasenhall. J.W.G.

3. C. *Clinopodium*, Benth. Hedge C. (Wild-Basil.) July—September.
Hilly, bushy pastures, hedgebanks, etc., frequent.

VII. NEPETA. Nepeta.

1. N. *Glechoma*, Benth. Ground-Ivy N. (Glechoma hederacea.) March—May.

Abundant everywhere.

2. N. *Cataria*, L. Catmint N. July—September.

Hedges about Bury, Bungay, Freston, Brandeston, Great Glemham and Easton.

VIII. PRUNELLA. Prunella.

1. P. *vulgaris*, L. Common P. (Self-heal.) July, August.

Common everywhere.

IX. SCUTELLARIA. Skullcap.

1. S. *galericulata*, L. Common S. July, August.

Sides of ditches and wet places about Bury, Brettenham Park, Great Glemham, Framlingham, Benhall Green, etc.

2. S. *minor*, L. Lesser S. July—October.

Very rare. Tuddenham. N.B.G.

XI. MARRUBIUM. Horehound.

1. M. *vulgare*, L. Common H. (White H.) Aug., September.

Waste ground, roadsides, etc., about Culford, Risby, Semer, Framlingham, etc.

XII. STACHYS. Stachys.

1. S. *Betonica*, Benth. Betony S. (Betonica officinalis.) June—August.

Woods, pastures, etc., at Rougham, Bradfield, Hitcham, Great Glemham, Lackford, and Barrow.

3. S. *sylvatica*, L. Hedge S. July, August.

Common in shady places, borders of woods, etc.

4. S. *palustris*, L. Marsh S. July, August.

Frequent on the banks of ditches, etc.

5. S. *arvensis*, L. Field S. April—November.

Fields about Hitcham, Great Glemham, Naughton, Felixstow, Bungay, etc.

XIII. GALEOPSIS. Galeopsis.

1. G. *Ladanum*, L. Red G. July—October.

Plentiful in corn fields at Hardwick, Barton, Rougham, Pakenham, Hitcham, Great Glemham, Mildenhall, Buigh Castle, etc.

3. G. *Tetrahit*, L. Common G. (Hemp-Nettle.) July—September.

Common in cultivated and waste ground.

Var. versicolor, About Bungay, Blyborough, and Sotterley. —Leiston, Yarmouth, Gillingham, and Beccles. O.B.G.

XIV. BALLOTA. Ballota.

1. B. *nigra*, L. Black B. (Black-Horehound.) June—October.

Hedges and waste places, abundant.

XV. LEONURUS. Leonurus.

1. L.* *Cardiaca*, L. Motherwort L. July—September.

About North Cove and Bungay. O.B.G.—Belton. Hist. Y.

XVI. LAMIUM. Lamium.

1. L. *amplexicaule*, L. Henbit L. April—Aug.

Plentiful in fields about Bury. Hitcham, Great Glemham, and Bungay.

2. L. *purpureum*, L. Red L. April—October.

Very common everywhere. *Var. incisum*, About Bury, Hitcham, Bungay, Great Glemham, etc.

3. L. *album*, L. White L. (Dead-Nettle.) May —September.

Abundant everywhere.

5. L. *Galeobdolon*, Crantz. Yellow L. (Archangel.) (Galeobdolon luteum.) April—June.

Woods and hedges. Hardwick, Hitcham, Bungay, Great Glemham, etc.

XVII. TEUCRIUM. Germander.

1. T. *Scorodonia*, L. Wood G. (Wood-Sage.) July, August.

Not uncommon in dry heathy ground at Thurston, Hitcham, Iken, Benhall, Bungay, Framlingham, West Stow, Tuddenham, etc.

2. T. *Scordium*, L. Water G. July, August.

Lakenheath—very rare.

XVIII. AJUGA. Bugle.

1. A. *reptans*, L. Creeping B. May, June.

Common in moist meadows, woods, etc.

ORD. LVIII. VERBENACEÆ.—VERBENANTHS, THE VERVEIN FAMILY.

I. VERBENA. Vervein.

1. V. *officinalis*, L. Common V. July—Sept.

Roadsides and waste ground, frequent.

ORD. LIX. PLUMBAGINEÆ.—PLUMBAGINANTHS, THE PLUMBAGO FAMILY.

I. STATICE. Statice.

1. S. *Limonium*, L. Common S. (Sea-Lavender.) July—September.

Salt marshes, frequent. Felixstow, Aldborough, Wherstead, Ipswich, etc. *Var. rariflora*, Hog Island, Orwell. Phyt.

II. ARMERIA. Thrift.

1. A. *vulgaris*, Willd. Common T. (Sea-Pink. Sea-Gilliflower.) May—September.

Not uncommon in salt marshes. Lowestoft, Aldborough, Wherstead, Orford, Bawdsey Ferry, etc.

ORD. LX. PLANTAGINEÆ.—PLANTAGINANTHS, THE PLANTAIN FAMILY.

I. PLANTAGO. Plantain.

1. P. *major*, L. Greater P. June—August.

Abundant on roadsides and waste places.

2. P. *media*, L. Hoary P. June—October.
Common in pastures in most districts.

3. P. *lanceolata*, L. Ribwort P. June, July.
Abundant everywhere.

4. P. *maritima*, L. Sea P. June—September.
Frequent in salt marshes about Yarmouth, Aldborough, Snape, Felixstow, etc.

5. P. *Coronopus*, L. Buckshorn P. June, July.
Dry gravelly soils, plentiful.

II. LITTORELLA. Littorel.

1. L. *lacustris*, L. Common L. June.
Wet places on Tuddenham heath.—Margin of Oulton Broad; Belton Common; Benacre and Cavenham. O.B.G.

SUB-CLASS IV. MONOCHLAMYDEÆ.—MONO-CHLAMYDS OR INCOMPLETE.

ORD. LXI. CHENOPODIACEÆ.—CHENOPODIANTHS, THE GOOSEFOOT FAMILY.

I. SALICORNIA. Salicorn.

1. S. *herbacea*, L. Common S. (Glasswort.)
August, September.
Common in muddy sea-wastes. *Var. radicans*, At Orford.

II. SUÆDA. Suæda. (Sea-Blite.)

1. S. *fruticosa*, Forsk. Shrubby S. (Salsola.) July—October.
Very rare. Near Walberswick Ferry.—Aldborough. J.W.G.

2. S. *maritima*, Dumort. Herbaceous S. (Chenopodium.) July—September.
Common on the sea-coast.

III. SALSOLA. Saltwort.

1. S. *Kali*, L. Prickly S. July.
Sandy sea-shores, common.

IV. CHENOPODIUM. Goosefoot.

1. **C. *Vulvaria*, L.** Stinking G. (C. olidum.)
August, September.
Very common under walls and in waste ground near the sea.

2. **C. *polyspermum*, L.** Many-seeded G. August,
September.
Frequent at Hitcham, Lavenham, Naughton, Great Glemham, Bergholt and Bungay.

3. **C. *album*, L.** White G. July—September.
Abundant in cultivated and waste ground.
Var. ficifolium, In gardens at Bury and Lakenheath.—
Waste places by the river. Hist. Y.

5. **C. *rubrum*, L.** Red G. August, September.
Waste ground at Lowestoft, and on the top of the cliff at Corton. Hitcham and Bungay.
Var. botryoides, Near the Yarmouth Railway Station.

6. **C. *urbicum*, L.** Upright G. August, Sept.
Not very common. Damp waste ground about Bury and Fornham. Plentiful on the margin of Rougham Canal. Bergholt and Bungay.—Corton. Eng. Fl.

7. **C. *murale*, L.** Nettle-leaved G. August, Sept.
Waste ground, frequent. About Yarmouth, Aldborough, Felixstow, Bungay, Slaughden Vale, etc.

8. **C. *hybridum*, L.** Maple-leaved G. August.
Gardens and cultivated ground at Bury, Mildenhall, Ipswich, and Naughton.—Sibton Abbey and Yarmouth. O.B.G.

9. **C. *Bonus-Henricus*, L.** Perennial G. (Good-King-Henry.) May—August.
Waste ground and way sides, frequent.

V. BETA. Beet.

1. **B. *maritima*, L.** Common B. June—Sept.
Not unfrequent along the coast. About Yarmouth, Lowestoft, Aldborough, Wherstead, Ipswich, and Walton ferry.

VI. ATRIPLEX. Orache.

1. **A. *portulacoides*, L.** Purslane O. (Sea-Purslane.) August—October.
Muddy sea-wastes, common.

2. A. *pedunculata*, L. Stalked O. July, August.

Salt marshes and waste ground about Braydon Broad and Aldborough.

4. A. *patula*, L. Common O. June—October.

The four principal varieties of this exceedingly variable plant are common in cultivated and waste ground, and their maritime forms on the coast.

5. A. *rosea*, L. Frosted O. (A. laciniata.) July —September.

Not unfrequent on the sandy coast from Yarmouth to Landguard Fort.

ORD. LXII. POLYGONACEÆ.—POLYGONANTHS, THE POLYGONUM FAMILY.

I. RUMEX. Dock.

1. R. *crispus*, L. Curled D. June—August.

Waysides and waste ground, common.

3. R. *obtusifolius*, L. Broad D. July—Sept.

Abundant everywhere.

4. R. *Hydrolapathum*, Huds. Water D. July, August.

Sides of rivers and ditches, frequent.

5. R. *conglomeratus*, Murr. Clustered D. June —August.

Watery places, common.

6. R. *sanguineus*, L. Red-veined D. July.

Damp waste ground about Bury.—Lowestoft and Framlingham. O.B.G.

Var. viridis, Plentiful in woods and shady places.

7. R. *pulcher*, L. Fiddle D. June—August.

Common on roadsides and waste ground.

8. R. *maritimus*, L. Golden D. July, August.

Marshes and waste ground about Yarmouth, Lowestoft and Bungay.—Dunwich ; Kessingland Dam, and near Hengrave. O.B.G.

9. R. *Acetosa*, L. Sorrel D. May—July.

Abundant in moist meadows and pastures.

10. R. *Acetosella*, L. Sheep-Sorrel D. May—July.

Plentiful in a dry barren soil.

III. POLYGONUM. Polygonum.

1. P. *aviculare*, L. Knotweed P. (Knotgrass.) May—September.

In waste and cultivated ground, abundant.

3. P. *Convolvulus*, L. Climbing P. (Climbing-Buckwheat.) July—September.

Plentiful in corn fields, etc.

6. P. *Bistorta*, L. Bistort P. (Snakeweed.) June —September.

Moist meadows and pastures in various places. Near the Pest-houses, Bury. Near Ollands House, Bungay. Near Wetherden Church. Woolpit, Hitcham, Sudbury, Halesworth, and Framlingham. Small plantation in Worlingham park.—Near Rendham Parsonage and Shipmeadow. O.B.G.

7. P. *amphibium*, L. Amphibious P. July, Aug.

Margins of ponds, ditches, and damp ground, frequent. Bury, Hitcham, Great Glemham, Benhall green, Framlingham, Bungay, Ipswich, etc.

8. P. *Persicaria*, L. Persicaria P. July—Oct.

Common in cultivated and waste ground.

9. P. *lapathifolium*, L. Pale P. July, August.

With the preceding, but not so common.

10. P. *Hydropiper*, L. Waterpepper P. Aug., Sep.

Ditches and watery places, frequent. Bury, Lackford, Hitcham, Great Glemham, Benhall, Bungay, etc.

11. P. *minus*, Huds. Slender P. Aug., Sept.

Rare. About Bungay.

ORD. LXIII. THYMELEACEÆ.—THYMELANTHS,
THE DAPHNE FAMILY.

I. DAPHNE. Daphne.

1. D.* *Mezereum*, L. Mezereon D. Feb—April.
Recorded from Bergholt.
2. D. *Laureola*, L. Spurge D. (Spurge-Laurel.)
January—May.
Frequent in woods and hedges in a clayey soil. Hitcham,
Thorpe, Great Glemham, Bergholt, etc.

ORD. LXIV. ELÆAGNACEÆ.—ELÆAGNANTHS,
THE ELÆAGNUS FAMILY.

I. HIPPOPHAE. Hippophae.

1. H. *rhamnoides*, L. Common H. (Sallow-Thorn.
Sea-Buckthorn.) May—July.
Very rare. Thorpe, near Aldborough.

ORD. LXV. SANTALACEÆ.—SANTALANTHS, THE
SANDALWOOD FAMILY.

I. THESIUM. Thesium.

1. T. *linophyllum*, L. Flax-leaved T. (Bastard-
Toadflax.) May—July.
Rare. Brandon.—Chalk bank, near the plantation of firs,
on Risby heath. O.B.G.

ORD. LXVI. ARISTOLOCHIACEÆ.—ARISTOLOCHI-
ANTHS, THE ARISTOLOCHIA FAMILY.

I. ARISTOLOCHIA. Birthwort.

1. A.* *Clematitis*, L. Common B. June—Sept.
Naturalized among old ruins, stony ground, etc. Abbey
Ruins, Bury.—Stuston, near Diss. O.B.G.

H

ORD. LXVII. EUPHORBIACEÆ.—EUPHORBI-
ANTHS, THE SPURGE FAMILY.

I. EUPHORBIA. Spurge.

2. E. *Helioscopia*, L. Sun S. June—October.
Common in cultivated ground.

3. E. *platyphylla*, L. Broad S. June—October.
Suffolk. Br. Fl.

6. E. *Peplus*, L. Petty S. July—November.
Abundant in cultivated and waste ground.

7. E. *exigua*, L. Dwarf S. July—October.
Fields and waste places, frequent.

8. E.* *Lathyrus*, L. Caper S. June, July.
Sparingly in the lane between Ickworth park and Saxham
gate.

10. E. *Paralias*, L. Sea S. August—November.
Sea coast, rather scarce. About Landguard Fort and Walton Ferry, Dunwich, Minsmere Level, and Thorpe.

12. E. *amygdaloides*, L. Wood S. April, May.
Frequent in woods and hedges in a heavy soil. Hardwick,
Bradfield, Chedburgh, Hitcham, Great Glemham, Berg-
holt, Bungay, &c.

II. MERCURIALIS. Mercury.

1. M. *perennis*, L. Perennial M. (Dog's Mercury.)
March—May.
Abundant in woods and shady places.

2. M. *annua*, L. Annual M. July—November.
Plentiful in gardens and waste ground about Bury and
Yarmouth. Also at Ipswich and Framlingham.

ORD. LXIX. CALLITRICHINEÆ.—CALLITRI-
CHANTHS, THE CALLITRICHE FAMILY.

I. CERATOPHYLLUM. Ceratophyll.

1. C. *demersum*, L. Common C. (Hornwort.)
July.

Ditches and slow streams, frequent.
Var. submersum, Wangford by Lakenheath.—Not uncommon about Yarmouth and Gorleston. Hist. Y.

II. CALLITRICHE. Callitriche.

1. C. *aquatica,* Sm. Common C. April—Sept.

A very variable plant; its usual forms are abundant in ditches throughout the county : the pedunculate variety grows about Ipswich.

ORD. LXX. URTICACEÆ.—URTICANTHS, THE NETTLE FAMILY.

I. URTICA. Nettle.

I. U. *urens,* L. Small N. June—September.

Cultivated and waste ground, abundant.

2. U. *pilulifera,* L. Roman N. June—August.

Waste ground, especially in the neighbourhood of the sea. By the side of the fish-houses, North Denes, Lowestoft, plentiful. About Thorpe, Aldborough and Euston.—Gorleston. O.B.G.

3. U. *dioica,* L. Common N. June—September.

Very common on roadsides and waste ground.

II. PARIETARIA. Pellitory.

1. P. *officinalis,* L. Wall P. June—September.

Old walls at Bury, Fornham, Great Glemham, Framlingham, etc.

III. HUMULUS. Hop.

1. H. *Lupulus,* L. Common H. July, August.

Frequent in hedges and thickets.

ORD. LXXI. ULMACEÆ.—ULMANTHS, THE ELM FAMILY.

I. ULMUS. Elm.

1. U. *montana,* Sm. Wych E. March, April.

Woods and hedges, frequent.

2. U. *campestris*, Sm. Common E. March—May.

With the former, but more abundant.

ORD. LXXII. AMENTACEÆ.—AMENTIFERS OR
CORYLANTHS, THE CATKIN FAMILY.

I. MYRICA. Gale.

1. M. *Gale*, L. Sweet G. May—July.

Rare. Marshes directly south of Oulton Broad and Dyke,
and probably in other marshes in the neighbourhood.

II. ALNUS. Alder.

1. A. *glutinosa*, L. Common A. March, April.

Damp woods and banks of streams, frequent.

III. BETULA. Birch.

1. B. *alba*, L. Common B. April, May.

Frequent in woods.

IV. CARPINUS. Hornbeam.

1. C. *Betulus*, L. Common H. May.

Woods and hedges about Bury, Hardwick, Ickworth,
Hitcham, Bergholt, Bungay, etc.

V. CORYLUS. Hazel.

1. C. *Avellana*, L. Common H. (Hazel-Nut.)
February—April.

Plentiful in woods and hedges.

VI. FAGUS. Beech.

1. F. *sylvatica*, L. Common B. May.

Woods, etc., common.

VII. QUERCUS. Oak.

1. Q. *Robur*, L. British O. April, May.

Abundant in woods and hedges.

VIII. SALIX. Willow.

1. S. *pentandra*, L. Bay W. May, June.

Near Bungay, frequent. O.B.G.

2. S. *fragilis*, L.   Crack W.   April, May.
Mermaid's Pits, Bury.

3. S. *alba*, L.   Common W.   May.
Sides of streams and ditches, abundant.

4. S. *amygdalina*, L.   Almond W.   (Triandrous W.)
April—June.
Var. triandra, Common in osier-grounds, etc., at Fornham, Rougham, etc.
Var. lanceolata, At Bradwell.

5. S. *purpurea*, L.   Purple W.   March—May.
Var. rubra, Icklingham. O.B.G.
Var. Forbyana, Bradwell.
Var. Lambertiana, Lackford Bridge and Bradwell. O.B.G.

6. S. *viminalis*, L.   Osier W.   April, May.
Common in osier-grounds and wet places.
Var. stipularis, About Bury. O.B.G.

7. S. *Capræa*, L.   Sallow W.   April, May.
Common in woods and hedges.

8. S. *aurita*, L.   Round-leaved W.   April, May.
Frequent in woods and hedges.

9. S. *repens*, L.   Creeping W.   April, May.

A very variable species. *Var. fusca* is plentiful in boggy ground on Tuddenham and other heaths.

IX. POPULUS.   Poplar.]

1. P. *alba*, L.   White P.   (Abele.)   March, April.
Sparingly in moist woods about Bury, Hitcham, etc.

2. P. *tremula*, L.   Aspen P.   March, April.
Woods and thickets about Rougham, Hitcham, Great Glemham, etc.

3. P.* *nigra*, L.   Black P.   April.
Plentiful on river banks.

DIV. II. GYMNOSPERMÆ.—GYMOSPERMS.

ORD. LXXIII. CONIFERÆ.—CONIFERS OR PINANTHS, THE PINE FAMILY.

I. PINUS. Pine.

1. P. *sylvestris*, L. Scotch P. (Scotch-Fir.) May, June.

Frequent on heaths and moors.

II. JUNIPERUS. Juniper.

1. J. *communis*, L. Common J. May.

Not unfrequent in woods and heathy ground.

III. TAXUS. Yew.

1. T. *baccata*, L. Common Y. March.

Woods, etc., occasionally.

CLASS II.

MONOCOTYLEDONES. MONOCOTYLEDONS.

DIV. I. PETALOIDEÆ.—PETALOID.
ORD. LXXIV. TYPHACEÆ.—TYPHANTHS, THE BULRUSH FAMILY.

I. TYPHA. Bulrush.

1. T. *latifolia*, L. Great B. (Reedmace. Cat's-tail.) July, August.
Abundant on the margins of broads, dykes, ponds, etc.

2. T. *angustifolia*, L. Lesser B. July.
Very general, but less abundant than the great B.

II. SPARGANIUM. Sparganium.

1. S. *ramosum*, Huds. Branched S. (Bur-reed.) July.
In ditches and streams, common.

2. S. *simplex*, Huds. Simple S. July.
Frequent, in the same situations as the preceding.

3. S. *natans*, L. Floating S. July.
Not nearly so general as the two former. Stagnant waters in a boggy soil. About Gedding and Lavenham. Bricet wood. Mildenhall and Wangford.

ORD. LXXV. AROIDEÆ.—ARANTHS, THE ARUM FAMILY.

I. ARUM. Arum.

1. A. *maculatum*, L. Common A. (Cuckoo-pint. Wake-Robin. Lords-and-Ladies.) April, May.
Woods, thickets and hedges, plentiful.

II. ACORUS. Acorus.

1. A. *calamus*, L. Sweet A. (Sweet-Flag. Sweet-Sedge.) June.

Rare. River-side below Halesworth; by the Waveney between Beccles and Gillingham; ditches near the river at Belton and ditches at Burgh Castle. O.B.G.

ORD. LXXVI. LEMNACEÆ.—PISTIANTHS, THE DUCK-WEED FAMILY.

I. LEMNA. Duckweed.

1. L. *trisulca*, L. Ivy-leaved D. June, July.

Common in ditches and still waters throughout the county.

2. L. *minor*, L. Lesser D. July.

Abundant on the surface of ditches and ponds, everywhere.

3. L. *gibba*, L. Gibbous D. June—September.

Not so common as the two foregoing, but found in ponds and ditches at Bury, Great Glemham, Benhall, Stowmarket and Bungay.——Abundant about Yarmouth. Hist. Y.

4. L. *polyrrhiza*, L. Greater D. Flowers unknown in Britain.

Very common throughout the county, usually accompanied by L. *minor*.

ORD. LXXVII. NAIADEÆ.—NAIADANTHS, THE NAIAD FAMILY.

I. ZOSTERA. Zostera.

1. Z. *marina*, L. Common Z. (Grass-wrack.) July, August.

Abundant on the coast, at and below low water mark; in salt lakes, and in tidal rivers near their junction with the sea.

III. ZANNICHELLIA. Zannichellia.

1. Z. *palustris*, L. Common Z. (Horned-Pondweed.) May—August.

Common in ponds and ditches.

IV. RUPPIA. Ruppia.

1. R. *maritima*, L. Sea R. July, August.

Frequent on the coast in salt water ditches, etc., from Yarmouth to Landguard Fort.

V. POTAMOGETON. Pondweed.

1. P. *natans*, L. Broad P. June, July.

Abundant in ponds and stagnant water.
Var. plantagineus, In boggy ditches near the river at Tuddenham, and about Bungay.

2. P. *heterophyllus*, Schreb. Various-leaved P. June, July.

At Lakenheath.

3. P. *lucens*, Shining P. June, July.

About Bungay, in the Waveney.

4. P. *prælongus*, Wulf. Long P. July.

With the former, abundant.

5. P. *perfoliatus*, L. Perfoliate P. July.

About Bungay. Probably not uncommon.

6. P. *crispus*, L. Curly P. June, July.

Common in ditches, ponds, etc.

7. P. *densus*, L. Opposite P. June, July.

Abundant in ditches in most districts.

8. P. *pusillus*, L. Slender P. July.

A very variable species. *Var. pusillus*, occurs at Tuddenham, Hitcham, St. Olave's bridge, Framlingham, Bungay, etc. *Var. gramineus*, at Leiston, Lound, etc.

9. P. *pectinatus*, L. Fennel P. (P. filiformis.) June, July.

Probably a common plant. Sea-wastes about Aldborough, and Dunwich. Also at Bungay.

ORD. LXXVIII. ALISMACEÆ.—ALISMANTHS, THE ALISMA FAMILY.

I. BUTOMUS. Butome.

1. B. *umbellatus*, L. Common B. (Flowering-Rush.) June, July.

Sparingly in the Lark, at the back of the gaol, Bury.
Stowlangtoft, Mildenhall, Bramford, Ipswich, Framlingham, Sternfield, and Langham bridge.—Plentiful in the
river at Shelland, near Stowmarket.

II. SAGITTARIA. Arrowhead.

1. S. *sagittifolia*, L. Common A. July—Sept.

Plentiful in most districts in ditches and shallow streams.

III. ALISMA. Alisma.

1. A. *Plantago*, L. Common A. (Water-Plantain.)
June—August.

Common in the same situations as the preceding.

2. A. *ranunculoides*, L. Lesser A. May—Sept.

Sparingly in boggy ditches, etc., at Tuddenham, Somerleyton and Bungay.——Lound, Hopton and elsewhere
about Yarmouth; in Livermere Park; near Lackford
Bridge and Framlingham. O.B.G.

IV. DAMASONIUM. Damasonium.

1. D. *stellatum*, Pers. Star D. (Actinocarpus.)
June, July.

Very rare. Framlingham. Hist. F.

VI. TRIGLOCHIN. Triglochin.

1. T. *palustre*, L. Marsh T. (Arrow-grass.) June
—August.

Common in marshes and wet meadows.

2. T. *maritimum*, L. Sea T. May—September.

Common in salt marshes and sea wastes.

ORD. LXXIX. HYDROCHARIDEÆ.—HYDROCHA-
RIDANTHS, THE HYDROCHARIS FAMILY.

I. ELODEA. Elodea.

1. E.* *canadensis*, Rich. Canadian E. (Anacharis
alsinastrum.) July—October.

Frequent in rivers and ditches. Abundant in the Lark
and in neighbouring ditches at Fornham, Hengrave,
Lackford, Santon Downham, etc. Also at Hitcham.

II. HYDROCHARIS. Frogbit.

1. H. *Morsus-ranæ*, L. Common F. July, Aug.

Common in ditches and ponds in most districts.

III. STRATIOTES. Stratiotes.

1. S. *aloides*, L. Water S. (Water-Soldier.) July.

Ditches in fenny districts, frequent. Abundant at Somerleyton, Haddiscoe, etc. Brandon, Bungay and Worlingham.—Loudham Hall mere; Bradwell and elsewhere
about Yarmouth. O.B.G.

ORD. LXXX. ORCHIDACEÆ.—ORCHIDANTHS,
THE ORCHID FAMILY.

I. MALAXIS. Malaxis.

1. M. *paludosa*, Sw. Bog M. (Bog-Orchis.)
July—Sept.

Abundant on Ashby Warren. Belton Common. N.B.G.
Probably not so rare as is generally supposed, but owing
to its small stature difficult to find.

II. LIPARIS. Liparis.

1. L. *Loeselii*, Rich. Two-leaved L. (Sturmia.)
July.

Very scarce. Bogs at Lakenheath. Bogs on Tuddenham heath, very sparingly.

IV. EPIPACTIS. Epipactis.

1. E. *latifolia*, Sw. Broad E. July, August.

Woods at Flixton and Fakenham.

3. E. *palustris*, Sw. Marsh E. July.

Boggy and marshy ground at Tuddenham, Lackford, Lowestoft and Bungay.

VI. LISTERA. Listera.

1. L. *ovata*, Br. Tway-blade L. May—July.

Frequent in woods and moist pastures.

VII. NEOTTIA. Neottia.

1. N. *Nidus-avis*, L. Birds'-nest N. (Listera.)
May, June.

Damp, shady woods at Rushbrook, Stowlangtoft, Hitcham,
Great Glemham and Bungay. Brakey Grove, Drinkstone. Chediston Hall wood. Little Saxham, Parham and Onehouse. O.B.G.

IX. SPIRANTHES. Spiranth.

1. S. *autumnalis*, Rich. Common L. (Lady's-
Tresses.) (Neottia spiralis.) August, September.

Damp pastures, etc., at Ickworth, Mildenhall, Groton,
Euston, Great Glemham and Bungay.

II. ORCHIS. Orchis.

1. O. *Morio*, L. Green-winged O. June.

Common in meadows and pastures.

3. O. *ustulata*, L. Dwarf O. May, June.

Hilly pastures, rare. Chalk bank near the chalk pit on
Risby heath. Devil's Dyke, Newmarket heath.

4. O. *mascula*, L. Early O. April, May.

Moist meadows and woods, common.

6. O. *maculata*, L. Spotted O. June, July.

Frequent in open woods, pastures, etc., Dr. White finds
an unusual and very handsome variety in old pastures
round Felsham Hall wood, flowering a little later than
the common form.

7. O. *latifolia*, L. Marsh O. June, July.

Marshes, common.

8. O. *hircina*, Scop. Lizard O. July.

Very scarce, if not extinct. One specimen was found at
Great Glemham in 1847, by the Rev. E. N. Blomfield,
who has since searched for it in vain.

9. O. *pyramidalis*, L. Pyramidal O. July, Aug.

Frequent in dry woods, open fields, etc. about Bury,
Felsham, Hitcham, Great Glemham, and Bungay.

10. O. *conopsea*, L. Fragrant O. (Gymnadenia.)
June—August.

Damp pastures at Hardwick, Hawstead, Hitcham, Naughton, Bungay and Framlingham. Abundant in bogs
about Mildenhall.

XII. HABENARIA. Habenaria.

1. H. *bifolia*, Br. Butterfly H. June—August.

Not uncommon in woods and damp pastures. Bury, Barton, Hawstead, Rushbrook, Pakenham, Hitcham, Bungay and Great Glemham.

3. H. *viridis*, Br. Green H. June—August.

Damp pastures, etc., frequent. Hawstead, Ickworth, Pakenham, Great Welnetham, Hitcham, Great Glemham, Little Thurlow, Bungay and Wangford.—-Cransford, Mettingham, Yoxford, Bradwell, Halesworth and Harleston. O.B.G.

XIII. ACERAS. Aceras.

1. A. *anthropophora*, Br. Man A. (Man-Orchis.) June.

Woods and pastures. Rushbrook, Rougham, Drinkstone, Hitcham, Stanton, Sternfield and Bungay. Chalk pit at Dallinghoe.—Old gravel pit near Sudbury; Little Saxham; Hawstead and Blakenham. O.B.G.

XIV. HERMINIUM. Herminium.

1. H. *monorchis*, Br. Musk H. (Musk-Orchis.) June, July.

Very rare. Chalk pit near Sicklesmere, and at Little Saxham. O.B.G.

XV. OPHRYS. Ophrys.

1. O. *apifera*, Huds. Bee O. (Bee-Orchis.) June, July.

Pastures and damp open woods, frequent. Hardwick, Ickworth, Drinkstone, Stowlangtoft, Hitcham, Great Glemham, Naughton, Bungay, Otley, Oulton, Flixton, and Baylham.—Yoxford, Harleston, and Brampton. O.B.G.

2. O. *aranifera*, Huds. Spider O. (Spider-Orchis.) April, May.

Rare. Pastures about Ickworth, Chevington, and Kennett. Chalk pit at Dallinghoe.—Westley Bottom; Great and Little Saxham; and near Sicklesmere. O.B.G.

3. O. *muscifera*, Huds. Fly O. (Fly-Orchis.) May—July.

Not unfrequent in pastures and woods. Chevington, Pakenham, Thurston, Hitcham, Baylham, Naughton, and Bungay. Chalk pit at Dallinghoe.

ORD LXXXI. IRIDEÆ.—IRIDANTHS, THE IRIS
FAMILY.

I. IRIS. Iris.

1. I. *Pseudacorus*, L. Yellow I. (Yellow-Flag.)
May—August.

Sides of ditches, marshes and watery places. Mermaid's
Pits, Bury. Hawstead Vale, Hitcham, Great Glemham,
Oulton, Somerleyton, Tuddenham, etc.

2. I. *fœtidissima*, L. Fetid I. (Roast-beef-plant.)
June, July.

Sparingly at Hitcham and Great Glemham. Naughton,
Offton and Bungay.——Blyborough, and between Parham and Easton. O.B.G.

ORD. LXXXII. AMARYLLIDEÆ.—AMARYLLI-
DANTHS, THE AMARYLLIS FAMILY.

I. NARCISSUS. Narcissus.

1. N. *Pseudo-Narcissus*, L. Daffodil N. (Daffy-
down-dilly.) March, April.

Pastures at Horringer. Pakenham Wood. Hitcham and
Saxmundham.——Herringfleet. Hist. Y.

II. GALANTHUS. Snowdrop.

1. G.* *nivalis*, L. Common S. February, March.

Ubbeston.—Pastures at Yoxford, Middleton and Westleton, and hedges at Laxfield. O.B.G.

III. LEUCOIUM. Snowflake.

1. L.* *æstivum*, L. Summer S. May.

Plentiful in pastures at Little Stonham. O.B.G.

ORD. LXXXIII. DIOSCORIDEÆ.—DIOSCOREANTHS,
THE YAM FAMILY.

I. TAMUS. Tamus.

1. T. *communis*, L. Common T. (Black-Bryony.)
May—July.

Hedges and thickets, common.

ORD. LXXXIV. LILIACEÆ.—LILIANTHS, THE LILY FAMILY.

I. PARIS. Paris.

1. P. *quadrifolia*, L. Common P. (Herb-Paris.) May, June.

Not unfrequent in woods and shady places. Link woods, Rushbrook. Saxham, Pakenham, Stowlangtoft, Hitcham, Bungay, etc.

II. POLYGONATUM. Solomon-Seal.

1. P. *multiflorum*, All. Common S. (Convallaria.) May, June.

In the Lily pits at Bradwell. O.B.G.—Gorleston. Eng. Fl.

III. CONVALLARIA. Convallaria.

1. C. *majalis*, L. Sweet C. (Lily-of-the-Valley.) May, June.

Abundant in Woolpit wood, near Haughley park.

IV. ASPARAGUS. Asparagus.

1. A. *officinalis*, L. Common A. June—August.

Said to be plentiful on the coast in some places.

V. RUSCUS. Ruscus.

1. R. *aculeatus*, L. Common R. (Butcher's-Broom.)

Rare. Snape and Bentley.—Near Lowestoft. With.

VI. FRITILLARIA. Fritillary.

1. F. *Meleagris*, L. Common F. (Snake's-head.) April.

Moist meadows and pastures.—Hawstead, Stratford and Bergholt.—Mickfield. N.B.G.—About Laxfield and Little Stonham. O.B.G.

VII. TULIPA. Tulip.

3. T. *sylvestris*, L. Wild T. April.

Very scarce. Reported to grow at Rougham. Formerly in the old chalk pits near St. Peter's Barn, Bury.

IX. GAGEA. Gagea.

1. G. *lutea*, Ker. Yellow G. (Ornithogalum. March—May.

Very rare. Great Glemham. Waller's Grove, near Ipswich.

X. ORNITHOGALUM. Ornithogalum.

1. O.* *umbellatum*, L. Common O. (Star-of-Bethlehem.) May, June.

In a pasture near St. Peter's Barn, Bury. Lakenheath, Mildenhall and Hitcham.—Hopton Churchyard, Little Saxham and Barton. O.B.G.

2. O.* *nutans*, L. Drooping O. April, May.

With the preceding near St. Peter's Barn. Wetherden, Bergholt and Sudbury.—Middleton and Framlingham. O.B.G.

XI. SCILLA. Squill.

1. S. *nutans*, Sm. Bluebell S. (Agraphis. Hyacinthus.) April—June.

Damp woods, plentiful.

XII. MUSCARI. Muscari.

1. M.* *racemosum*, Mill. Grape M. (Grape-Hyacinth.) May.

Sparingly in Mr. Lee's woods, Bury. Abundant at Lakenheath. Plentiful by the road-side near Higham Station. Pakenham.—Fields at Hengrave and plantations at Cavenham. O.B.G.

XIII. ALLIUM. Allium.

3. A. *oleraceum*, L. Field A. (Field Garlic.) July.
Rare. Lakenheath.

6. A. *vineale*, L. Crow A. (Crow Garlic.) June.

Pasture near St. Peter's Barn, Bury. Hitcham.—Near the Orwell, common. Ips. Fl.

7. A. *ursinum*, L. Broad A. (Ramsons.) April —June.

Woods and shady places. Hitcham and Great Glemham. Felsham Hall Wood.

XVII. COLCHICUM. Colchicum.

1. **C. *autumnale*, L.** Common C. (Meadow-Saffron.) August—October.

Not unfrequent in moist meadows and pastures, varying occasionally with white flowers. Hawstead in abundance. Drinkstone Green. North of Barton Church. All Saints, Hitcham, Bergholt, Rumburgh, Benhall, Bildeston, Debach and Cretingham.—Little Saxham, Laxfield, Soham and Easton. O.B.G.

ORD. LXXXV. JUNCACEÆ.—JUNCANTHS, THE RUSH FAMILY.

I. JUNCUS. Rush.

1. **J. *communis*, Mey.** Common R. July.

The two extreme forms J. *conglomeratus* and J. *effusus*, with many intermediate ones are plentiful in marshes and wet places.

2. **J. *glaucus*, Ehrh.** Hard R. July.

Common by road-sides and in wet pastures.

5. **J. *articulatus*, L.** Jointed R. (J. lamprocarpus, acutiflorus, etc.) June—August.

An extremely variable species. Its leading forms are very generally distributed throughout the county.

6. **J. *obtusiflorus*, Ehrh.** Obtuse R. August.

Frequent in marshes and fenny situations about Bury, Fornham, Hitcham, Great Glemham, Snape and Bungay. Pakenham Fen.—Common about Yarmouth. Hist. Y.

7. **J. *compressus*, Jacq.** Round-fruited R. June—August.

Marshy ground, especially near the sea. Lowestoft, Aldborough, Dunwich, Snape, and Hitcham.

8. **J. *squarrosus*, L.** Heath R. June, July.

Frequent in a heathy, moory soil, about West Stow, Tuddenham, Snape, Ipswich, Bungay, &c.

9. **J. *bufonius*, L.** Toad R. August.

Common in watery or muddy places.

10. J. *capitatus,* Weig. Capitate R. (J. supinus.) May—July.

Very rare. Snape.

11. J. *maritimus,* Lam. Sea R. July, August.

Not unfrequent in marshy places on the coast. Walton Ferry, Snape, Dunwich, about Braydon Broad, etc.

12. J. *acutus,* L. Sharp R. July.

Salt marshes at Aldborough.

II. LUZULA. Woodrush.

1. L. *pilosa,* Willd. Hairy W. March—May.

Woods and banks at Hitcham, Great Glemham, Rougham, Bungay, etc.

4. L. *campestris,* Br. Field W. April, May.

Common in dry pastures and heathy ground.

DIV. II. GLUMACEÆ.—GLUMACEOUS.

ORD. LXXXVII. CYPERACEÆ.—SEDGES OR CYPERANTHS, THE SEDGE FAMILY.

II. SCHÆNUS. Schænus.

1. S. *nigricans,* L. Black S. (Bog-rush.) June, July.

Bogs at Tuddenham and Mildenhall. Pakenham Fen. About Bungay.

III. CLADIUM. Cladium.

1. C. *Mariscus,* Br. Prickly C. (Twig-rush.) (Schænus.) July, August.

Bogs at Lakenheath, Mildenhall and Tuddenham. Pakenham Fen.—Near Mutford Bridge and in marshes between Bungay and Beccles. O.B.G.

IV. RHYNCHOSPORA. Beaksedge.

2. R. *alba,* Vahl. White B. (Schænus.) June—August.

Abundant in bogs at Belton and Lound. O.B.G.

V. BLYSMUS. Blysmus.

1. B. *compressus*, Panz. Broad B. (Schænus.) June, July.

Bogs and marshes. Tuddendam, Aldborough, Dunwich and Bungay. Pakenham Fen and Benhall Green.—Flixton and Middleton. O.B.G.

VI. SCIRPUS. Scirpus.

1. S. *acicularis*, L. Needle S. (Eleocharis.) July, August.

Belton Common. O.B.G. Probably not unfrequent.

2. S. *palustris*, Creeping S. (Eleocharis.) June, July.

Common in marshy places, shallow ditches, etc.

3. S. *multicaulis*, Sm. Many-stalked S. (Eleocharis.) July.

Worlingham.—Abundant in bogs on Belton Common. O.B.G.—Common about Yarmouth. Hist. Y.

4. S. *pauciflorus*, Lightf. Few-flowered S. July, August.

Rare. Moory ground at Mildenhall, Snape, Worlingham and Bungay.—About Lound. Hist. Y.—Belton and Bradwell Commons. O.B.G.

5. S. *cæspitosus*, L. Tufted S. June, July.

Frequent in bogs and moist heathy places, though we have few recorded localities. Lowestoft and Westleton.

6. S. *fluitans*, Floating S. (Isolepis.) June, July.

Rare. Mildenhall.

7. S. *setaceus*, L. Bristle S. (Isolepis.) July, August.

Hitcham and Bungay, and most likely not uncommon elsewhere.

8. S. *Savii*, Seb. et Maur. Savi's S. (Isolepis.) July.

Scarce. Recorded from Snape and Benhall.

12. **S. *lacustris*, L.** Lake S. July, August.

Abundant in ditches and watery places. The glaucous
variety is also very general.

13. **S. *maritimus*, L.** Sea S. July, August.

Frequent in watery places in the neighbourhood of the sea.
Yarmouth, Lowestoft, Somerleyton, Sizewell, Snape,
Walton, etc.

14. **S. *sylvaticus*, L.** Wood S. July.

Rare. Great Glemham.—Low meadows by the river at
Rendham; Cransford, Sweffling and Yoxford. O.B.G.

VII. ERIOPHORUM. Cottonsedge.

3. **E. *polystachyum*, L.** Common C. May, June.

Var. angustifolium, Common in marshes.
Var. latifolium, Marshes about Somerleyton.

IX. CAREX. Carex. (Sedge.)

1. **C. *dioica*, L.** Diæcious C. May, June.

Not very general. Bogs at Mildenhall, Wangford, Tuddenham, Hopton and Bungay.

2. **C. *pulicaris*, L.** Flea C. May, June.

Bogs on Tuddenham heath, Great Glemham, Bungay, etc.

5. **C. *leporina*, L.** Oval C. (C. ovalis.) June.

Moist meadows and watery places. Hitcham, Great Glemham, Sizewell, Bungay, Ipswich, Tuddenham, etc.

8. **C. *stellulata*, Good.** Star-headed C. May, June.

Not uncommon in marshy ground. Snape, Benhall
Green, Bungay, Mildenhall, Tuddenham, etc.

9. **C. *canescens*, L.** Whitish C. (C. curta.) June.

Bogs and marshes, rare. Snape.—Westleton and Flixton.
O.B G.—Framlingham. Hist. F.

10. **C. *remota*, L.** Remote C. June.

Plentiful in moist and shady places.

11. **C. *axillaris*, Good.** Axillary C. June.

Occasionally and sparingly. Watery places at Hardwick,
Hitcham, Chediston and Mettingham.—Badingham and
Cransford. J.W.G.—Metfield. O.B.G.

12. C. *paniculata*, L. Panicled C. June.

Frequent in a boggy soil. Mermaid's Pits Bury, plentiful.
Sides of ditches at Icklingham. Mildenhall, Sudbourn
and Bungay.
Var. teretiuscula, Tuddenham, Mildenhall and Bungay.
Fritton decoy, and elsewhere about Yarmouth. Salt
marshes at Dunwich.—Darsham and Bradwell. O.B.G.

13. C. *vulpina*, L. Fox C. June.

Very common by the sides of ditches, ponds, etc.

14. C. *muricata*, L. Prickly C. May, June.

Gravelly pastures and marshy ground, common.
Var. divulsa, Hardwick, Hitcham, Great Glemham,
Bungay, Middleton, etc.

15. C. *arenaria*, L. Sand C. June.

Abundant on the sandy coast, and inland at West Stow,
Tuddenham and Mildenhall.
Var. intermedia, in marshy places. Mermaid's Pits, Bury.
Tuddenham, Mildenhall, Great Glemham, Bungay and
Sizewell.—Somerleyton. N.B.G.—Belton. Hist. Y.—By
the Gipping. Ips. Fl.

16. C. *divisa*, Huds. Divided C. May, June.

Rare. Salt marshes at Ipswich.—Bawdsey. W.L.N.

19. C. *cæspitosa*, L. Tufted C. (C. vulgaris.)
May, June.

Vars. b. and c. frequent. Deep bogs near the river at
Tuddenham, Snape, Great Glemham, Hitcham, Bungay,
Ipswich, etc.

20. C. *acuta*, L. Acute C. May.

Not uncommon about Bungay and elsewhere.

26. C. *precox*, Jacq. Vernal C. April, May.

Abundant in dry pastures and heaths.

28. C. *pilulifera*, L. Pill-headed C. June.

Not very general. Tuddenham heath. Snape. Otley
wood. Westleton, etc.

30. C. *filiforme*, L. Slender C. May.

Rare. About Mildenhall, Brandon, Lakenheath and
Worlingham.—Bogs at Lound. O.B.G.

31. C. *hirta*, L.   Hairy C.   May, June.

Pastures, marshes, woods, etc., frequent

32.   C. *pallescens*, L.   Pale C.   June.

Damp woods and marshy places. Hitcham and Felsham Hall woods.   Great Glemham, Bungay, etc.

33.   C. *extensa*, Good.   Long-bracted C.   June.

Rare in marshes near the sea.   Near the river at Belton. —Aldborough. J.W.G.

34.   C. *flava*, L.   Yellow C.   May, June.

Bogs and marshy places frequent.   Hopton, Tuddenham, Lakenheath, Snape, Benhall Green and Bungay.
Var. *Œder'* occurs at Tuddenham, Belton, Lound, Bradwell, Bungay, Framlingham, etc.

35.   C. *distans*, L.   Distant C.   June.

Not uncommon in marshy situations about Lowestoft, Aldborough, Dunwich, etc.
Var. fulva, Tuddenham heath; Mildenhall and Bradwell. —Theberton Common. O.B.G.
Var. binervis, About Bungay.—Mildenhall. J.T.—Corton and other heaths about Yarmouth. O.B.G.

37.   C. *panicea*, L.   Carnation C.   (Carnation-Grass.) June.

Frequent in moist meadows, marshes, etc.

39.   C. *limosa*, L.   Mud C.   June.

Very scarce.   Belton bog. N.B.G.

40.   C. *glauca* Scop.   Glaucous C.   June.

Plentiful in a variety of situations.

41.   C. *sylvatica*, Huds.   Wood C.   May, June.

Common in woods at Hardwick, Felsham, Hitcham, Great Glemham, Bungay, etc.

42.   C. *strigosa*, Huds.   Thin-spiked C.   May, June.

Very rare.   Brent Eleigh, occasionally.

43.   C. *Pseudocyperus*, L.   Cyperus-like C.   June.

Rather local.   Watery places at Hardwick, Felsham, Hitcham, Great Glemham, Bungay, Somerleyton, Bradwell and Wangford.—Badingham. J.W.G.

44. C. *pendula*, Huds. Pendulous C. May, June.

Frequent in a clayey soil. Hardwick, Hawstead, etc. Roadside from Bradfield to Hitcham, plentiful.—Near Woodbridge. O.B.G.

45. C. *ampullacea*, Good. Bottle C. June.

Bogs at West Stow, Tuddenham, Mildenhall, Wangford, Bungay, Snape, Farnham and Great Glemham.—Westleton, Yoxford, Bradwell and Lound. O.B.G.

46. C. *vesicaria*, L. Bladder C. May, June.

Rare. Boggy pool in Hitcham wood.—Badingham. J.W.G.

47. C. *paludosa*, Good. Marsh C. May.

The two varieties of this plant *C. paludosa* and *C. riparia*, are abundant in watery places.

ORD. LXXXVIII. GRAMINEÆ.—GRASSES OR FESTUCANTHS, THE GRASS FAMILY.

II. MILIUM. Milium. (Millet-Grass.)

1. M. *effusum*, L. Spreading M. June.

Not unfrequent in moist woods and shady places. Hardwick, Gedding, Felsham, Hitcham, Great Glemham, etc.

III. PANICUM. Panicum.

1. P. *sanguinale*, L. Fingered P. (Digitaria.) July, August.

Very plentiful in the sand lands at Sutton and near Henham.—Fields opposite the Ferry at Woodbridge. O.B.G.

5. P. *viride*, L. Green P. (Setaria.) July, Aug.

Lakenheath. With the preceding at Sutton and Woodbridge.—Near Wickham Market and Bungay. O.B.G.

V. ANTHOXANTHUM. Anthoxanth.

1. A. *odoratum*, L. Sweet A. (Vernal-Grass.) May, June.

Pastures and woods, plentiful.

VII. DIGRAPHIS. Digraphis.

1. D. *arundinacea*, Trin. Reed D. (Phalaris.) July, August.

River banks and shallow ditches, common. The variegated variety occurs at Lakenheath.

VIII. PHLEUM. Phleum.

1. P. *pratense*, L. Timothy P. (Cat's-tail.) June.

Abundant everywhere.

3. P. *Bœhmeri*, Schrad. Bœhmer's P. (Phalaris phleoides.) July.

Mildenhall and Kentford. Formerly on Haberden, Bury.

5. P. *arenarium*, L. Sand P. (Phalaris.) May, June.

Sandy ground near the sea and in the north-west. Thorpe, Felixstow, Walton and Sudbourn. Newmarket, Mildenhall, Tuddenham, Icklingham and West Stow. Old walls at Bury, rare.

IX. ALOPECURUS. Foxtail.

1. A. *agrestis*, L. Slender F. June, July.

Very general in cultivated and waste ground—often a troublesome weed in corn fields.

2. A. *pratensis*, L. Meadow F. May, June.

Abundant in meadows and pastures.

3. A. *geniculatus*, L. Marsh F. July, August.

Frequent in wet and marshy places.
Var. fulvus is said to grow in a pond, near the road, half way between Bury and Sicklesmere.
Var. bulbosus, Salt marshes by Braydon Broad. N.B.G. —Abundant in marshes by river side at Belton and Burgh Castle. O.B.G.

XIII. AGROSTIS. Agrostis. (Bent-Grass.)

1. A. *alba*, L. Common A. (A. vulgaris et alba.) July, August.

Abundant everywhere.

2. A. *canina*, L. Brown A. June, July.

Moist heathy ground at Tuddenham, Snape, etc.

4. A. *Spica-venti*, L. Silky A. (Apera.) June, July.

> Rare. Fields and sandy ground at Lakenheath, Milden-hall, Tuddenham, Cavenham, Boyton, Snape, Great Glemham and Felixstow. Blaxhall and Bromswell. O.B.G.
>
> *Var. interrupta*, Waste ground near Bury, Higham, and Tuddenham.

XV. PSAMMA. Maram.

1. P. *arenaria*, Beauv. Sea M. (Sea-Matweed.) (Ammophila arundinacea.) July.

> Abundant on the sandy coast, where it is invaluable in binding together the otherwise shifting soil.

XVI. CALAMAGROSTIS. Smallreed.

1. C. *Epigeios*, Roth. Wood S. (Arundo.) July.

> Woods, thickets, and hedges, frequent. Hardwick, Hit-cham, Mildenhall, Drinkstone and Bungay. In the Hyde at Westley.

2. C. *lanceolata*, Roth. Purple S. (Arundo Cala-magrostis.) June.

> Very local, but abundant in fenny plantations about Mildenhall and Lakenheath. Hitcham and Naughton woods.

XVII. AIRA. Aira. (Hair-grass.)

1. A. *cæspitosa*, L. Tufted A. June, July.

> Moist places, borders of fields, etc., plentiful.

2. A. *flexuosa*, L. Wavy A. July.

> This is either very rare in the county, or it seldom flowers. Recorded only from the neighbourhood of Ipswich "occasionally."

3. A. *canescens*, L. Grey A. (Corynephorus.) July.

> Very rare. Lowestoft Denes.

4. A. *precox*, L. Early A. May, June.

> Abundant in hilly and sandy pastures.

5. A. *caryophyllea*, L. Silvery A. June, July.

Plentiful in a light, heathy soil.

XVIII. AVENA. Oat.

1. A. *fatua*, L. Wild O. June—August.

Corn fields and waste places, frequent.
Var. strigosa, Waste ground about Bury.

2. A. *pratensis*, L. Perennial O. June, July.

Probably not a common plant, but plentiful on borders of fields and road-sides at Hardwick, and also on Risby heath. Said to be common about Ipswich.
Var. pubescens is equally frequent in dry pastures all round Bury.

3. A. *flavescens*, L. Yellow O. (Trisetum.) July.

Common everywhere.

XIX. ARRHENATHERUM. False-Oat.

1. A. *avenaceum*, Beauv. Common F. (Holcus.) June, July.

Plentiful in various situations.

XX. HOLCUS. Holcus.

1. H. *lanatus*, L. Common H. June, July.

Abundant everywhere.

2. H. *mollis*, L. Soft H. July.

In damp or shady situations, not very frequent nor plentiful. Bury, Hardwick, Hitcham, Great Glemham, Bungay, Lowestoft, Santon Downham, etc.

XXII. SPARTINA. Spartina.

1. S. *stricta*, Sm. Cord S. (Cord-Grass.) (Dactylis.) August.

Salt marshes, Slaughden Vale.—Aldborough and Orford, abundantly O.B.G.

XXIII. Lepturus. Lepturus.

1. L. *incurvatus*, Trin. Curved L. (Rottboellia.) July.

Waste ground near Lake Lothing, Lowestoft. Sizewell and Snape.

XXIV. NARDUS. Nard.

1. N. *stricta*, L. Common N. (Matgrass.) June.

Plentiful in a heathy soil. Lowestoft, West Stow, Tuddenham, Mildenhall, Snape, Dunwich, Bungay, etc.

XXV. ELYMUS. Lymegrass.

1. E. *arenarius*, L. Sand L. July.

Sandy sea-shores, sparingly. Pakefield cliff, flowering freely. Southwold, Thorpe, and Easton Bavents.

XXVI. HORDEUM. Barley.

2. H. *pratense*, Huds. Meadow B. June, July.

Meadows and pastures not unfrequent. Fornham, Hardwick, Hitcham, Great Glemham, Bungay, etc.

3. H. *murinum*, L. Wall B. June, July.

Abundant in waste places, under walls, etc.

4. H. *maritimum*, With. Sea B. June.

Frequent in salt marshes and sea-wastes. Abundant about Braydon Broad. Aldborough, Snape, Orford, Felixstow, Catwade Bridge, etc.

XXVII. TRITICUM. Triticum.

1. T. *repens*, L. Couch T. (Couchgrass.) June —August.

Abundant in cultivated and waste ground, hedges, etc. *Var. junceum* is very plentiful on the sandy coast.

2. T. *caninum*, Huds. Fibrous T. July.

Moist woods and shady places. Hardwick, Hawstead, Offton, Dunwich, etc.

XXVIII. LOLIUM. Lolium.

1. **L. *perenne*, L.** Ryegrass L. June, July.
Abundant everywhere.
*Var.** *italicum,* Frequent in cultivated fields.

2. **L. *temulentum*, L.** Darnel L. July.
Not uncommon in corn fields and waste ground. Bury,
Hitcham, Great Glemham, Bungay, etc.
Var. arvense, About Bury, Hitcham, Ipswich & Bungay.

XXIX. BRACHYPODIUM. False-Brome.

1. **B. *sylvaticum*, Beauv.** Slender F. (Bromus.)
June, July.
Frequent in woods and hedges, except where the soil is
sandy.

2. **B. *pinnatum*, Beauv.** Heath F. (Bromus.) July.
Very rare. About Bungay. O.B.G.

XXX. BROMUS. Brome.

1. **B. *erectus*, Huds.** Upright B. June, July.
Very rare. Dunwich. J.W.G.—Mildenhall. Phyt.

2. **B. *asper*, L.** Hairy B. June, July.
Borders of woods and moist hedges, plentiful.

3. **B. *sterilis*, L.** Barren B. June.
Abundant everywhere.

6. **B. *arvensis*, L.** Field B. June—August.
Plentiful among Saintfoin, etc. all round Bury, evidently
introduced with the seed.
Var. secalinus, Corn fields south of Lake Lothing and
about Bungay.—Corn fields about Framlingham. Hist. F.
Var. mollis, Abundant everywhere.
Var. racemosus, Frequent in meadows, pastures, etc.

7. **B *giganteus*, L.** Tall B. (Festuca.) July,
August.
Woods and shady places, not very common. Hitcham,
Great Glemham, Ampton, Bergholt, Bungay, Santon
Downham, etc.
Var. triflora, Hardwick and elsewhere.

XXXII. FESTUCA. Fescue.

1. **F. *ovina*, L.** Sheep's F. June, July.
   a. *ovina*, Common S. F. Abundant in dry pastures.
   b. *duriuscula*, Tall S. F. Common in moister situations.
   c. *rubra*, Sand S. F. Sandy ground, especially near the sea. Lowestoft and Sizewell.

2. **F. *pratensis*, L.** Meadow F. June, July.
   a. *loliacea*, Spiked M. F. Not unfrequent in pastures.
   b. *pratensis*, Common M. F. Common in pastures, etc.
   c. *elatior*, Tall M. F. Moist rich ground about streams, etc. Hitcham. Oulton Dyke. Mermaid's Pits, Bury.

4. **F. *Myurus*, L.** Rat's-tail F. (F. bromoides, etc.) June.

   Frequent in waste ground, wall tops, chalk pits, etc. Common about Bury. West Stow, Cavenham, Tuddenham, Offton, Mildenhall and Great Glemham.—About Snape and Aldborough. O.B.G.
   Var. pseudo-Myurus, At Snape.

5. **F. *uniglumis*, Soland.** One-glumed-F. June.

   Very scarce. About Landguard Fort. W.L.N.

XXXII. DACTYLIS. Cock's-foot.

1. **D. *glomerata*, L.** Clustered C. June, July.

   Abundant everywhere.

XXXIII. CYNOSURUS. Dog's-tail.

1. **C. *cristatus*, L.** Crested D. July.

   Abundant in dry pastures, etc.

XXXIV. BRIZA. Quakegrass.

1. **B. *media*, L.** Common Q. June.

   Frequent in meadows, pastures and marshes.

XXXV. POA. Poa.

1. **P. *aquatica*, L.** Reed P. (Glyceria.) July, August.

Frequent in ditches and shallow streams about Bury, Lowestoft, Somerleyton, Bungay, and various other places.

2. P. *fluitans*, Scop. Floating P. (Glyceria.) July, August.

Common in ditches and watery places.

3. P. *maritima*, Huds. Sea P. (Sclerochloa.) July.

Salt marshes and sea-wastes, frequent all along the coast.

4. P. *distans*, L. Reflexed P. (Sclerochloa.) July, August.

Waste ground near Lake Lothing Lowestoft, frequent. Also about Yarmouth.

5. P. *procumbens*, Curt. Procumbent P. (Sclerochloa.) June, July.

Common in roads in the marshes about Yarmouth.—— Cobholm. N.B.G.

6. P. *rigida*, L. Hard P. (Sclerochloa.) June.

Walls and dry waste ground, common.

7. P. *loliacea*, Huds. Darnel P. (Triticum.) June, July.

Sandy sea-shores. Aldborough and Dunwich.—Lowestoft, Gorleston, Southwold and Landguard Fort. O.B.G.

8. P. *annua*, L. Annual P. April—October.

The commonest of all grasses.

9. P. *compressa*, L. Flattened P. June, July.

Walls and barren ground and sides of fields at Bury, Hitcham, etc.

10. P. *pratensis*, L. Meadow P. June, July.

Plentiful in pastures, etc.

11. P. *trivialis*, L. Roughish P. June, July.

Equally common with the former in moister or more shady situations.

12. P. *nemoralis*, L. Wood P. June, July.

Apparently very local. Woods about Felsham and Bungay.

15. P. *bulbosa*, L. Bulbous P. April, May.

Lowestoft Denes.—Sandy ground at Aldborough. O.B.G.

XXXVI. CATABROSA. Catabrose.

1. C. *aquatica*, Beauv. Water C. (Aira.) May, June.

Wet ditches and watery places at Mildenhall, Hengrave, Great Glemham and Bungay. Near Flatford Mill.

XXXVII. MOLINIA. Molinia.

1. M. *cærulea*, Mœnch. Purple M. (Melica.) July, August.

Moory ground, plentiful. Tuddenham, West Stow, Lowestoft, Great Glemham, Westleton, etc. Pakenham Fen.

XXXVIII. MELICA. Melick.

2. M. *uniflora*, L. Wood M. May—July.

Shady woods at Hardwick, Felsham, Gedding, Hitcham, Great Glemham, Bungay, Bergholt, etc.

XXXIX. TRIODIA. Triodia.

1. T. *decumbens*, Beauv. Decumbent T. (Heathgrass.) July.

Frequent in a heathy soil. Hitcham, rare. Snape, Lowestoft, Corton, Westleton, Great Glemham, Bungay, West Stow, Tuddenham, etc.

XL. KÆLERIA. Kæleria.

1. K. *cristata*, Pers. Crested K. (Aira.) June, July.

Dry pastures, banks, etc. Plentiful at Bury, Ampton, Mildenhall, Tuddenham, etc. About Bungay.

XLII. ARUNDO. Reed.

1. A. *Phragmites*, L. Common R. (Phragmites communis.) July, August.

Abundant in broads, marshes, ditches, etc.

CLASS III.

ACOTYLEDONES.—ACOTYLEDONS OR CRYPTOGAMS.

ORD. LXXXIX. LYCOPODIACEÆ.—THE CLUBMOSS FAMILY.

PILULARIA *globulifera*, L. Wet places on Hopton Common and at Bungay.
LYCOPODIUM *clavatum*, L. Tuddenham heath, sparingly.
—— *inundatum*, L. Bogs on Tuddenham and Westleton heaths. Lound & Belton Commons. Herringfleet North Border.

ORD. XC. EQUISETACEÆ.—THE EQUISETUM OR HORSETAIL FAMILY.

EQUISETUM *Telmateia*, Ehrh. Marshy and watery places at Ipswich, Ramsholt, Sutton, Bungay, Lowestoft, etc.
—— *sylvaticum*, L. Suffolk. Moore.
—— *arvense*, L. Common in damp ground in most districts.
—— *limosum*, L. Frequent in marshy places. Langham, Hitcham, Great Glemham, etc.
—— *palustre*, L. Common in a boggy soil.
Var. polystachyon, At Tuddenham.
—— *hyemale*, L. Marshy or boggy ground at Hardwick, the Bradfields, and Drinkstone. Ditch-sides at Hitcham.

ORD. XCI. FILICES.—THE FERN FAMILY.

OPHIOGLOSSUM *vulgatum*, L. Moist meadows and pastures at Little Welnetham, Rushbrooke, Hawstead, Ickworth, Saxham, Bardwell, Hitcham, Bungay, etc.

BOTRYCHIUM *Lunaria*, Sw. Pastures, bogs, etc. Rougham, Melton, Kennett, Lakenheath, Mildenhall, Blundeston, Fritton, Hopton, Browston, Leiston, etc.

OSMUNDA *regalis*, L. Fritton Decoy and Leiston.

POLYPODIUM *vulgare*, L. Common in old hedgebanks, & trees.

ASPIDIUM *aculeatum*, Sw. Hedgebanks and shady places about Bury, Bungay, Hitcham, etc.

    Var. lobatum, Damp shady banks about Bury, Bradfield, Rushbrooke, Groton, Belton, Sotterley, etc.

    Var. angulare, Hitcham, etc.

—— *Thelypteris*, Sw. Boggy woods and marshy places. Wood near Icklingham. Cavenham Severals. Fritton Decoy. Bungay, Mildenhall, Wangford and Blundeston. Lound, Hopton, Belton and Bradwell Commons.

—— *Oreopteris*, Sw. Bradwell.

—— *Filix-mas*, Sw. Very general.

—— *cristatum*, Sw. Very rare. Fritton Decoy. Near the old Decoy at Westleton. Bexley Decoy, near Ipswich.

—— *spinulosum*, Sw. Great Glemham, etc.

    Var. dilatatum, Wangford, Benhall, Fritton, etc.

ASPLENIUM *Filix-fœmina*, Bernh. Frequent in moist woods, etc. Hitcham Wood. Fritton Decoy. Bungay, etc.

—— *Trichomanes*, L. Walls, banks, etc. Walls of Rushbrooke Hall. Hengrave, Bungay, etc. Benhall and Aldborough Churches.

—— *Adiantum-nigrum*, L. Frequent on hedgebanks, walls, etc. Saxham, Rougham, Hitcham, and Bungay. Parham Church.

—— *Ruta-muraria*, L. Old walls, etc. Elmswell, Culford, Boxtead and Bungay. Rushbrooke Hall Moat. Lowestoft, Aldborough, Framlingham and Saxmundham Churches.

SCOLOPENDRIUM *vulgare*, Sm. Common on damp and shady banks, etc.

BLECHNUM *Spicant*, Roth. (B. boreale.) Fritton Decoy and Snape Bog.

PTERIS *aquilina*, L. Abundant in woods, heaths, etc.

CYSTOPTERIS *fragilis*, Bernh. Yoxford and Bungay. Moore.

ORD. XCII. MUSCI.—THE MOSS FAMILY.

PHASCUM *serratum*, Schreb. Shady banks, Bury & Bradwell.

—— *alternifolium*, Dicks. Moist banks at Barton, in fruit.

—— *subulatum*, L. Banks, etc. Rougham, Hawstead, etc.

—— *patens*, Hedw. Fields and banks. Wangford.

—— *muticum*, Schreb. Moist banks, not common.

PHASCUM *cuspidatum*, Schreb. Damp banks & fields, frequent.
—— *bryoides*, Dicks. Banks and fields, rare. Lakenheath.
SPHAGNUM *obtusifolium*, Ehrh. Bogs, common.
—— *squarrosum*, W. and M. Bogs, Belton and Somerley.
—— *acutifolium*, Ehrh. Bogs, very common.
—— *cuspidatum*, Ehrh. Bogs, frequent.
GYMNOSTOMUM *viridissimum*, Sm. Trunks of trees. Nowton, Barton Mills, Great Blakenham and Yarmouth.
—— *curvirostrum*, Hedw. Clay pit at Bradwell.
—— *ovatum*, Hedw. Common on banks.
—— *truncatulum*, Hoffm. Frequent in fields, banks, etc.
—— *Heimii*, Hedw. Banks. Wattisfield, Bradwell, etc.
—— *fasciculare*, Hedw. Belton Common.
—— *pyriforme*, Hedw. Ditch-banks, etc., About Bury.
—— *microstomum*, Hedw. Not unfrequent on banks.
SPLACHNUM *ampullaceum*, L. Bogs, etc. Belton, Tuddenham, Wangford.
ENCALYPTA *vulgaris*, Hedw. Banks. Nowton, Rougham, Horringer and Bungay.
WEISSIA *Starkeana*, Hedw. Fields near Bury. Clay pit at Belton. Ufferton.
—— *lanceolata*, H. and T. Frequent on banks.
—— *cirrata*, Hedw. Pales of Hengrave & Sotterley Parks.
—— *controversa*, Hedw. Frequent on banks.
—— *calcarea*, Hedw. Chalk pits about Bury, frequent.
GRIMMIA *apocarpa*, Hedw. Old walls and trees, frequent.
—— *pulvinata*, Sm. Abundant on wall tops.
DIDYMODON *purpureus*, H. & T. On the ground, plentiful.
—— *rigidulus*, Hedw. Near Bungay.
TRICHOSTOMUM *lanuginosum*, Hedw. Heaths at West Stow and Santon Downham. North Denes, Lowestoft.
DICRANUM *bryoides*, Sw. Common on moist banks.
—— *adiantioides*, Sw. Not unfrequent on moist banks.
—— *taxifolium*, Sw. With the former.
—— *glaucum*, Hedw. Tuddenham and other heaths.
—— *cerviculatum*, Hedw. Belton and Bradwell Commons.
—— *flexuosum*, Hedw. Belton Common.
—— *crispum*, Hedw. Bank near Herringfleet Hall.
—— *undulatum*, Ehrh. Heaths about Bury, frequent.
—— *scoparium*, Hedw. Woods and heaths, common. In fruit at West Stow.
—— *varium*, Hedw. Chalk pits and lanes about Bury. Bradwell, Eriswell, Weston, etc.
—— *heteromallum*, Hedw. Sandy shady banks, frequent.
TORTULA *enervis*, H. & G. Bradwell, Wangford & Eriswell.
—— *rigida*, Turn. Barton, Hawstead, Bildeston, etc.
—— *convoluta*, Sw. Waste ground at Thurston
—— *revoluta*, Brid. Ditto on Railway near Bury.
—— *muralis*, Hedw. Abundant everywhere.

TORTULA *ruralis*, Sw.   Thatch, walls and trees, common.
—— *subulata*, Hedw.   Common on shady banks.
—— *unguiculata*, H. and T.   Very common.
—— *cuneifolia*, Turn.   Hopton and Belton Commons.
—— *fallax*, Sw.   Grassy banks at Bury, Barton, etc.
POLYTRICHUM *undulatum*, Hedw.   Woods and shady banks.
—— *piliferum*, Schreb.   Heaths, common.
—— *juniperum*, Willd.   Heaths, frequent.
—— *commune*, L.   Heathy places, common.
—— *aloides*, Hedw.   Moist shady banks and heaths.
—— *nanum*, Hedw.   With the preceding at Rougham, etc.
FUNARIA *hygrometrica*, Hedw.   In a variety of situations.
ORTHOTRICHUM *anomalum*, Hedw.   Icklingham churchyard
    wall.   Old walls at Bury.
—— *affine*, Schrad.   Common on old pales, trees, etc.
—— *diaphanum*, Schrad.   Trunks of trees about Bury,
    Wangford, Henham, Yarmouth, etc.
—— *striatum*, Hedw.   Trees about Bury, rare.
—— *crispum*, Hedw.   Trees at Bungay, Rendham, etc.
BRYUM *palustre*, Sw.   Bogs at Icklingham, Bungay, etc.
—— *dealbatum*, Dicks.   Suffolk.   Mr. Eagle.
—— *carneum*, L.   Sides of ditches at Bury and Bradwell.
    Ickworth park, near water.
—— *argenteum*, L.   Abundant on walls and waste places.
—— *pyriforme*, Sm.   Wangford.
—— *capillare*, L.   Banks, walls, etc., frequent.
—— *cæspititium*, L.   Abundant everywhere.
    Var. lacustre, at Wangford.
—— *turbinatum*, Sw.   Stony, heathy ground at Barton.
—— *nutans*, Schreb.   Belton Common and Ashby Warren.
—— *roseum*, Schreb.   Rougham heath.   About Bungay.
—— *ligulatum*, Schreb.   Woods and moist banks, common.
    In fruit in a wood at Barton.
—— *rostratum*, Schrad.   Great Glemham.
—— *cuspidatum*, Schreb.   Woods & moist banks, frequent.
—— *affine*, Brid.   Rare.   Woods at Fornham St. Martin
    and Mildendall, fruiting freely.
BARTRAMIA *pomiformis*, Hedw.   Suffolk.
—— *fontana*, Sw.   Bogs at Tuddenham, Icklingham, Bun-
    gay, etc.
LEUCODON *sciuroides*, Schwaegr.   Common on trees.
ANOMODON *curtipendulum*, H. & T.   Thatch of boat-house
    in Herringfleet Decoy.
—— *viticulosum*, H. & T.   Frequent on trees near the ground.
DALTONIA *heteromalla*, H. & T.   Trees.   Barton, Barton
    Mills, and Yarmouth.
FONTINALIS *antipyretica*, L.   Ditches in Ickworth Park.
    Little Saxham.

HYPNUM *trichomanoides,* L.   Trunks of trees, frequent.
—— *complanatum,* L.   Trunks of trees, common.
—— *riparium,* L.   Watery places, frequent.
—— *undulatum,* L.   Lakenheath.
—— *denticulatum,* L.   Not uncommon in woods, etc.
—— *medium,* Dicks.   Westleton, on trunks of trees.
—— *tenellum,* Dicks.   Walls of Framlingham Castle.
—— *serpens,* L.   Trunks of trees, etc., common.
—— *stramineum,* Dicks.   Belton Bog.   Near Browston Hall.
—— *purum,* L.   Abundant everywhere.
—— *piliferum,* Schreb.   Banks at Hardwick & Bradwell.
—— *Schreberi,* Willd.   Frequent on banks, heaths, etc.
—— *sericeum,* L.   Abundant on trees, etc.
—— *lutescens,* Huds.   Banks, plentiful.
—— *albicans,* Neck.   Heathy ground, Rougham & Bungay.
—— *alopecurum,* L.   Common on shady banks, etc.
—— *dendroides,* L.   Fornham, Rougham, etc.
—— *curvatum,* Swartz.   Woods, frequent about Bury.
—— *myosuroides,* L.   Bungay, on trees.
—— *splendens,* Hedw.   Heaths, hedge-banks, etc.
—— *proliferum,* L.   Woods and heathy places, common.
—— *prælongum,* L.   Abundant on banks, trees, etc.
—— *abietinum,* L.   Elden, Nowton, and Barton.
—— *rutabulum,* L.   Abundant on banks and trees.
—— *velutinum,* L.   With the former, abundant.
—— *ruscifolium,* Neck.   Bury, Hawstead, etc.
—— *striatum,* Schreb.   Frequent on trees, banks, etc.
—— *confertum,* Dicks.   Wangford.
—— *cuspidatum,* L.   Bogs and watery places, common.
—— *cordifolium,* Hedw.   Bogs.   Bradwell.   Belton and
       Herringfleet Commons.
—— *stellatum,* Schreb.   Marshes at Fornham St. Martin, etc.
—— *triquetrum,* L.   Woods and heaths, plentiful.
—— *squarrosum,* L.   Heaths, banks, etc., common.
—— *filicinum,* L.   Watery places.   Bury and Bungay.
—— *palustre,* L.   Barton Mills.
—— *fluitans,* L.   Watery places.
—— *aduncum,* L.   Bogs. Tuddenham, Wangford & Bungay.
—— *uncinatum,* Hedw.   Belton Common and Wangford.
—— *rugulosum,* W. & H.   Heathy places about Thetford.
—— *commutatum,* Hedw.   Wet places about Bury.
—— *scorpioides,* L.   Bog at Tuddenham.
—— *cupressiforme,* L.   Heaths, trunks of trees, etc., plen-
       tiful.
—— *molluscum,* Hedw.   About Bury and Bungay, rare.

ORD. XCIII. HEPATICÆ.—THE LIVERWORT
FAMILY.

RICCIA *crystallina*, L. Fallows, Hardwick, Wangford, etc.
—— *fluitans*, L. Ditches at Hopton. Hitcham.
—— *natans*, L. Stagnant pools. Hitcham Wood, Laken-
heath and Henley.
SPHÆROCARPUS *terrestris*, Sm. Fallows, etc., Hardwick,
Bungay, Burgh and Bradwell. Often accompanied
by Riccia crystallina.
ANTHOCEROS *punctatus*, L. Moist ground. Wangford and
Holton.
TARGIONIA *hypophylla*, L. Banks near Nayland.
MARCHANTIA *polymorpha*, L. Moist shady places, common.
—— *conica*, L. Santon Downham, Great Glemham, etc.
JUNGERMANNIA *asplenioides*, L. Moist woods, frequent.
—— *Sphagni*, Dicks. On Sphagnum. Belton Bogs.
—— *crenulata*, Sm. Moist heathy ground at Herringfleet, etc.
—— *bicuspidata*, L. Shaded heathy ground at Tuddenham, etc.
—— *byssacea*, Roth. Heathy ground at Wangford.
—— *connivens*, Dicks. Bogs on Gorleston Common.
—— *pusilla*, L. Herringfleet.
—— *nemorosa*, L. Woolpit and other woods.
—— *undulata*, L. Belton bogs.
—— *exsecta*, Schmid. Boggy ground at Tuddenham, etc.
—— *complanata*, L. Frequent on trunks of trees.
—— *scalaris*, Schrad. Ashby Warren.
—— *Trichomanis*, Dicks. Ashby Warren. Belton Bogs.
Gorleston Common.
—— *inflata*, Huds. Tuddenham heath.
—— *bidentata*, L. Hedgebanks, woods, etc., common.
—— *Francisci*, Hook. Suffolk.
—— *setacea*, Web. Near Ashby Decoy.
—— *platyphylla*, L. Abundant on trees, walls, etc.
—— *ciliaris*, L. Tuddenham and Lound Heaths.
—— *tomentella*, Ehrb. Suffolk.
—— *dilatata*, L. Frequent on trunks of trees.
—— *pinguis*, L. Suffolk.
—— *epiphylla*, L. Ditch banks at Santon Downham.
—— *furcata*, L. On trees and bushes. Hardwick, Great
Glemham, etc.

ORD. XCIV. LICHENES.—THE LICHEN FAMILY.

BÆOMYCES *roseus*, Pers. Heaths, occasionally.
CALICIUM *sessile*, Pers. On Porina pertusa, frequent.
—— *tigillare*, Pers. Rare, on pales and rails.
—— *tympanellum*, Ach. Gate-posts, etc., at Chediston,
Yarmouth, Halesworth and Wangford.

CALICIUM *ferrugineum,* T. and B. Old pales at Rumburgh.
—— *clavellum,* T. and B. Timber at Bury.
—— *hyperellum,* Ach. Old trees about Bury.
—— *chrysocephalum,* Ach. Park pales, Sotterley.
—— *phæocephalum,* T. and B. Barns at Bruisyard.
—— *chlorellum,* Ach. Scotch Firs about Bury.
—— *curtum,* T. and B. Decaying timber about Bury, etc.
—— *sphærocephalum,* Ach. Old wood and trees.
—— *æruginosum,* T. and B. Old pales near Bury.
ARTHONIA *impolita,* Borr. Oaks near Halesworth.
OPEGRAPHA *lyncea,* Borr. Trees about Bungay, Halesworth,
    Great Glemham and Herringfleet.
—— *atra,* Pers. Smooth bark, frequent.
—— *saxatilis,* D.C. On walls, etc., occasionally.
—— *scripta,* Ach. Smooth bark, frequent.
VERRUCARIA *nitida,* Schrad. Frequent on smooth bark.
—— *plumbea,* Ach. Chalk stones at Thetford.
—— *viridula,* Ach. Walls at Gorleston, etc.
—— *nigrescens,* Pers. Old walls about Yarmouth.
ENDOCARPON *Hedwigii,* Ach. Barrow Bottom.
—— *sorediatum,* Mud walls, Thetford.
—— *smaragdalum,* Ach. Thetford.
—— *fuscellum,* Churches about Bury, Bradwell & Gorleston.
PERTUSARIA *communis,* D.C. Abundant on trees.
—— *fallax,* Ach. Old oaks about Bury, Yarmouth, Hales-
    worth, Beccles and Harleston.
THELOTREMA *lepadinum,* Ach. Bark of trees, not uncommon.
LEPRARIA *viridis,* T. & B. Abundant on pales, trees, walls, etc.
—— *ochracea,* T. & B. Trees, Yarmouth and Halesworth.
—— *flava,* Ach. Frequent on rugged oaks, boards, etc.
—— *alba,* Ach. Common on trees, etc.
—— *virescens,* Sm. Trees in Shrubland Park.
SPILOMA *murale,* T. and B. Walls of Burgh Church.
—— *auratum,* Sm. Timber at Wrentham.
—— *nigrum,* T. and B. Old church walls and trees.
—— *fuliginosum,* T. and B. Rugged bark, Belton and
    Blundeston.
—— *decolorans,* T. and B. Old walls, etc., frequent.
VARIOLARIA *Vitiligo,* T. and B. Gate-posts, old rails, etc.
—— *conspurcata,* T. aud B. Churches about Yarmouth,
    Wangford and Halesworth. Burgh Castle.
—— *discoidea,* Pers. Bark of trees, etc., common.
—— *faginea,* Pers. Common on trees, pales, etc.
—— *aspergilla,* Ach. Pales at Wrentham, Herringfleet,
    Henham and Ickworth.
URCEOLARIA *gibbosa,* Ach. Flints at Mildenhall.
—— *cinerea,* Ach. Churches, etc., about Yarmouth and
    Lakenheath.
—— *rufescens,* Gorleston.

111

URCEOLARIA *calcarea*, Ach.   Stones and walls at Wangford.
—— *scruposa*, Ach.   Heaths, walls, etc., occasionally.
LECIDEA *scabrosa*, Ach.   Walls and tombs at Pakefield,
    Bradwell and Worlingham.
—— *Ehrhartiana*, Ach.   Old pales and barns at Livermere.
—— *albo-atra*, Borr.   Pakefield Church, Burgh Castle, and
    Wangford.
—— *confluens*, Ach.   Common on stones.
—— *prominula*, Borr.   Thetford, on flints.
—— *parasema*, Ach.   Smooth bark, common.
—— *Griffithii*, On Birch, Fritton and Ashby.   Sotterley
    Park pales.
—— *aromatica*, Turn.   Churches about Yarmouth and
    Bury.
—— *uliginosa*, Ach.   Bradwell and Lound heaths.
—— *abietina*, Ach.   Old oaks, Bury, Hopton & Herringfleet.
—— *Lightfootii*, Ach.   Sotterley Park pales.
—— *quadricolor*, Borr.   Heaths at Ashby, Dunwich, Lound
    and Herringfleet.
—— *coronata*, Borr.   Turfy ground about Yarmouth.
—— *anomala*, A.   Elms at Barton, Bradwell & Somerleyton.
—— *rupestris*, Ach.   Pakefield Church.
—— *vernalis*, Borr.   On Elms.   Hopton, Bradwell & Burgh.
—— *icmadophila*, Ach.   Heaths about Bury.
—— *marmorea*, Ach.   Ash trees at Stoven.
—— *atro-flava*, Turn.   Flints about Bury and Thetford.
—— *ulmicola*, Borr.   Old trees.   Link woods, Rushbrooke.
    Sotterley Park.   Shotford heath.
—— *aurantiaca*, Ach.   Trees at Ickworth and Haddiscoe.
LECANORA *atra*, Ach.   Walls, frequent.
—— *periclea*, Ach.   Old pales at Livermere.
—— *coarctata*, Ach.   Worlingham Church-yard walls.
—— *sophodes*, Ach.   Old trees.   Halesworth, Blundeston,
    and Ickworth.
—— *aspersa*, Borr.   Flints near Bury.
—— *subfusca*, Ach.   Smooth bark, abundant.
—— *cœsio-rufa*, Abundant on walls at Bury.
—— *Hœmatomma*, Ach.   Trees about Bury and Herring-
    fleet.   Walls at Gorleston, Geldeston & St. Olave's
    Bridge.
—— *cerina*, Ach.   Trees at Bradwell, Herringfleet, etc.
—— *chloroleuca*, Ach.   Walls of Burgh Castle.
—— *varia*, Ach.   Pales about Bury and Yarmouth.   Burgh
    Castle.
—— *albella*, Ach.   Common on smooth bark.
—— *Turneri*, Ach.   Sotterley park pales.   Herringfleet
    Hall.   Belton and Fritton Woods.
—— *vitellina*, Ach.   Pales, boards, etc., common.
PSORA *cœruleo-nigrescens*, Barrow and Wangford.

PSORA *scalaris*, Pales and trees. Halesworth, Sotterley, Herringfleet, Henham, Fornham, etc.
—— *decipiens*, Hoffm. Brandon.
SQUAMARIA *hypnorum*, Lound and West Stow heaths.
—— *cœsia*, Walls, etc. Burgh, Bradwell & Great Glemham.
—— *murorum*, Frequent on walls and stones.
—— *fulgens*, Wangford.
—— *lentigera*, D.C. Chalky heaths.
—— *circinnata*, On calcareous stones.
—— *elœina*, Trees about Yarmouth.
PLACODIUM *canescens*, D.C. Abundant everywhere, fruit rare.
PARMELIA *caperata*, Ach. Shrubland park.
—— *conspersa*, Ach. Stones on West Stow heath.
—— *scortea*, Ach. Trees, bricks, etc., Bury, Livermere, Melton and Framlingham.
—— *saxatilis*, Ach. On stones and trees, common.
—— *omphalodes*, Ach. Heaths about Bury.
—— *perlata*, Ach. Shrubland park.
—— *olivacea*, Ach. Common on trees, etc.
—— *corrugata*, Ach. Old thorns, Ickworth park. St. Peter's, near Bungay.
—— *pulverulenta*, Ach. Common on trees.
—— *pityræa*, Ach. Trees and walls, frequent.
—— *stellaris*, Ach. On trees, frequent.
—— *virella*, Ach. Trees about Yarmouth.
—— *incurva*, Stones on West Stow heath.
—— *aleurites*, Ach. Park pales at Ickworth. Ampton, Sotterley and Herringfleet.
—— *parietina*, Ach. Abundant everywhere.
—— *physodes*, Ach. Sotterley park pales, fruiting.
—— *cycloselis*, Ach. Trees about Yarmouth & Gisleham.
STICTA *pulmonaria*, Oaks in Barham Slade. Old trees about Easton and Parham, sparingly.
COLLEMA *nigrum*, Ach. Bradwell and Horringer Churches.
—— *microphyllum*, Ach. On elm bark, near Bury.
—— *cristatum*, Ach. On the ground about Bury.
—— *corrugatum*, Ach. Old trees, Ickworth, Great Welnetham and Beccles.
—— *cretaceum*, Ach. Chalk pits at Bury and Wangford.
—— *palmatum*, Sand hills at Corton.
—— *nigrescens*, Ach. Trunks of trees at Bury.
—— *granulatum*, Walls of Horringer Church.
—— *crispum*, Borr. Frequent about Bury.
—— *subtile*, Ach. Bury, Harleston, Lound, Belton, etc.
—— *tenuissimum*, Ach. Dry banks. Somerleyton, Burgh, Bungay, etc.
—— *sinuatum*, Burgh Castle. Walls at Bury.
—— *Schraderi*, Ach. Heaths about Bury.
PELTIDEA *canina*, Ach. Among moss, grass, etc., common.
—— *spuria*, Ach. Hedgebanks, occasionally.

CETRARIA *sepincola,* Ach. Worlingham park pales.
—— *glauca,* Ach. Heaths about Bury and Corton. Sotterley park pales.
BORRERA *ciliaris,* Ach. Trunks of trees, frequent.
—— *tenella,* Ach. Trees and bushes, frequent.
—— *furfuracea,* Ach. Ampton park pales.
EVERNIA *prunastri,* Ach. Trees & pales, common. In fruit on Sotterley Park pales, at Hengrave & Geldeston.
RAMALINA *fraxinea,* Ach. Abundant on trees.
—— *fastigiata,* Ach. Common on trees, pales, etc.
—— *farinacea,* Ach. Trees and pales, common. In fruit in Herringfleet Decoy, Gunton wood and Henham park.
—— *pollinaria,* Ach. Dead trees at Felixstow. Barn doors, frequent.
USNEA *plicata,* Ach. Hengrave park pales, in fruit.
ALECTORIA *jubata,* Ach. Brettenham, Sotterley and Worlingham park pales.
CORNICULARIA *aculeata,* Ach. Icklingham and Rougham heaths.
ISIDIUM *lutescens,* T. & B. Oaks, Ickworth park & Bramfield.
—— *coccodes,* Ach. Pales and bricks, Ickworth & Burgh.
CLADONIA *uncialis,* Heaths, frequent.
—— *rangiferina,* Hoffm. Heaths, abundant.
—— *furcata,* Hoffm. Heaths, frequent.
—— *caviosa,* Lound heath.
SCYPHOPHORUS *parasiticus,* On a decayed tree at Barham.
—— *alcicornis,* Heathy ground, Wangford & Icklingham.
—— *pyxidatus,* Heaths, etc., common.
—— *cocciferus,* Rougham and other heaths.
PYCNOTHELIA *Papillaria,* Somerley, Lound and Belton heaths.

ORD. XCV. CHARACEÆ.—THE CHARA OR NITELLA FAMILY.

CHARA *translucens,* Pers. Stagnant pools. Browston.
—— *polysperma,* Br. Running water at Bury.
—— *vulgaris,* L. Ditches, &c., not uncommon.
—— *hispida,* L. Ditches and ponds, frequent.

ORD. XCVI. ALGÆ. THE FAMILY OF SEAWEEDS, &c., &c.

SARGASSUM *vulgare,* Ag. Beach about Yarmouth, very rare.
CYSTOSEIRA *granulata,* Ag.   Ditto
—— *fibrosa,* Ag.   Ditto
HALIDRYS *siliquosa,* Lyng. Common on the coast.
FUCUS *vesiculosus,* L. Abundant on the coast.

FUCUS *serratus*, L. Felixstow, Walton, etc., common.
—— *nodosus*, L. Ditto.
—— *canaliculatus*, L. Ditto.
HIMANTHALIA *lorea*, Lyng. Felixstow, Yarmouth, &c.
LAMINARIA *digitata*, Lam. Felixstow.
—— *saccharina*, Lam. Common on marine rocks, &c.
DESMARESTIA *ligulata*, Lam. Corton and Gunton beaches.
SPOROCHNUS *pedunculatus*, Ag. Ditto.
—— *villosus*, Ag. Ditto.
DICTYOTA *dichotoma*, Lam. Ditto.
—— *atomaria*, Gr. Ditto.
CHORDARIA *flagelliformis*, Ag. Felixstow, &c.
CHORDA *Filum*, Lam. Ditto.
DICHLORIA *viridis*, Gr. Beach about Yarmouth.
CUTLERIA *multifida*, Gr. Ditto.
FURCELLARIA *fastigiata*, Lam. Ditto.
DELESSERIA *sanguinea*, Lam. Sea-shores, occasionally.
—— *sinuosa*, Lam. Corton and Gunton beaches.
—— *Hypoglossum*, Ag. Ditto.
—— *ruscifolia*, Lam. Ditto.
POLYIDES *rotundus*, Gr. Ditto.
NITOPHYLLUM *laceratum*, Gr. Ditto.
RHODOMENIA *bifida*, Gr. Ditto.
—— *laciniata*, Gr. Ditto.
—— *Palmetta*, Gr. Beach about Yarmouth.
PLOCAMIUM *coccineum*, Lyng. Common on the coast.
RHODOMELA *lycopodioides*, Ag. Beach near Yarmouth.
—— *subfusca*, Ag. Corton and Gunton beaches.
—— *scorpioides*, Ag. Ditto.
BONNEMAISONIA *asparagoides*, Ag. Ditto.
LAURENCIA *pinnatifida*, Lam. Rocks on the coast.
—— *dasyphylla*, Gr. Corton and Gunton beaches.
CHYLOCLADIA *clavellosa*, Ditto.
—— *ovalis*, Ditto.
GIGARTINA *confervoides*, Lam. Ditto.
HALYMENIA *ligulata*, Ag. Ditto.
—— *furcellata*, Ag. Ditto.
CHONDRUS *mammillosus*, Gr. Felixstow, etc.
—— *crispus*, Lyng. Ditto.
CHÆTOSPORA *Wigghii*, Ag. Beach about Yarmouth.
PORPHYRA *laciniata*, Ag. Abundant on marine rocks, etc.
ULVA *latissima*, L. Ditto.
—— *Lactuca*, L. Ditto.
—— *crispa*, Lightf. Under walls at Bury.
TETRASPORA *lubrica*, Ag. Corton and Gunton.
ENTEROMORPHA *compressa*, Gr. Common on the coast.
BRYOPSIS *plumosa*, Ag. Beach about Yarmouth, rare.
VAUCHERIA *Dillwynii*, Ag. On damp ground, common.
CLADOSTEPHUS *spongiosus*, Ag. Frequent on rocks in the sea.
SPHACELARIA *scoparia*, Lyng. Corton and Gunton beaches.

SPHACELARIA *cirrhosa*, Ag. Common on Fuci, etc.
ECTOCARPUS *littoralis*, Lyng. Ditto.
—— *siliculosus*, Lyng. Ditto.
—— *Mertensii*, Ag. Sea-shore about Yarmouth, rare.
POLYSIPHONIA *stricta*, Gr. Frequent on the coast.
—— *patens*, Gr. Felixstow.
—— *atro-rubescens*, Gr. Marine rocks, frequent.
—— *nigrescens*, Gr. Marine rocks, common.
—— *fastigiata*, Gr. Common on Fuci.
—— *elongata*, Gr. Frequent on the sea shore.
—— *byssoides*, Gr. Corton and Gunton beaches.
GRIFFITHSIA *equisetifolia*, Ag. Ditto.
—— *setacea*, Ag. Ditto.
CERAMIUM *rubrum*, Ag. Common on marine rocks, etc.
—— *ciliatum*, Ducl. Ditto.
CALITHAMNION *Turneri*, Ag. Beach about Yarmouth.
—— *roseum*, Ag. On plants in the river Yare.
—— *tetricum*, Ag. On marine rocks, common.
—— *fasciculatum*, Haw. Sea-shore about Yarmouth.
—— *Borreri*, Ag. Ditto.
—— *thuyoides*, Ag. Ditto.
—— *repens*, Lyng. Ditto.
—— *Rothii*, Lyng. Marine rocks, frequent.
—— *lanuginosum*, Lyng. On decaying Algæ, frequent.
BULBOCHÆTE *setigera*, Ag. On various plants in bogs at
        Tuddenham, Bradwell and Hopton.
CONFERVA *ericetorum*, Roth. Lound, Bradwell and other
        heaths.
—— *bombycina*, Ag. Hopton.
—— *zonata*, W. and M. On pebbles at Lound.
—— *vesicata*, Ag. Frequent in stagnant water.
—— *rivularis*, L. Streams and rivers, common.
—— *dissiliens*, Dill. Floodgate at Westleton.
—— *tortuosa*, Dill. About Braydon Broad, etc.
—— *flacca*, Dill. About Yarmouth.
—— *fucicola*, Vell. Common on Fuci.
—— *fracta*, Fl.Dan. Frequent in ditches.
    Var. β. In salt-water ditches, near Yarmouth.
—— *glomerata*, L. In the sea and fresh water, common.
—— *pellucida*, Huds. Corton and Gunton.
—— *rupestris*, L. Common on marine rocks, etc.
—— *riparia*, Roth. Salt pools about Yarmouth.
HYDRODICTYON *utriculatum*, Roth. About Bungay, etc.
MOUGEOTIA *genuflexa*, Ag. Clay pits about Yarmouth.
TYNDARIDEA *pectinata*, Haw. Pools on Bradwell heath, etc.
ZYGNEMA *nitidum*, Ag. Ditches about Yarmouth.
—— *decimimum*, Ag. Common in ditches.
—— *quininum*, Ag. Ditto.
CALOTHRIX *confervicola*, Ag. On marine Algæ, common.

CALOTHRIX *scopulorum*, Ag. Felixstow, etc., on marine rocks.
—— *distorta*, Ag. Ditches on Lound & Thetford heaths.
LYNGBIA *muralis*, Ag. Common on damp walls.
OSCILLATORIA *tenuis*, Ag. Frequent in muddy ditches.
—— *cyanea*, Ag. On damp walls inside Hengrave and
Icklingham Churches.
—— *decorticans*, Gr. Rotten timber, etc., frequent.
—— *nigra*, Vauch. Ditches, etc., common.
—— *autumnalis*, Ag. Abundant on damp walls.
CHROOLEPUS *aureus*, Common on trees, &c.
PROTONEMA *Acharii*, Ag. Damp shady banks, frequent.
—— *Orthotrichi*, Ag. Common on Orthotricha.
HYGROCROCIS *Atramenti*, Ag. Common on ink.
LEPTOMITUS *lacteus*, Ag. On stones in the Lark at Bury.
MESOGLOIA *vermicularis*, Ag. Sea-coast about Yarmouth, etc.
BATRACHOSPERMUM *moniliforme*, Ag. Ditches, rivulets, etc.,
common.
—————— *vagum*, Ag. About Ipswich.
DRAPARNALDIA *glomerata*, Ag. On stems of reeds in pools
at Bradwell, Hopton, Belton, etc.
—— *tenuis*, Ag. Sticks in rivulets, Browston & Thetford.
CHÆTOPHORA *endiviæfolia*, Hook. In running water, occasionally.
—— *tuberculosa*, Hook. Boggy pool at Santon Downham.
CORYNEPHORA *marina*, Ag. Common on Algæ and rocks.
RIVULARIA *atra*, Roth. Ditto.
—— *angulosa*, Roth. Peat bogs on Belton Common.
PALMELLA *cruenta*, Ag. Damp walls, very common.
—— *botryoides*, Lyng. Damp heathy ground, frequent.
NOSTOC *commune*, Vauch. Common in gravelly or sandy
places.
—— *pruniforme*, Ag. About Yarmouth.
—— *sphæricum*, Vauch. Ditches at Bradwell on Ceratophyllum demersum.
VOLVOX *globator*, Ponds at Bury and Ipswich.
CLOSTERIUM *Lunula*, Common in ditches, etc.
PEDIASTRIUM *Boryanum*, In the basin of a fountain at Bury.
SCENEDESMUS *quadricauda*, With the former, & at Ipswich.
EPITHEMIA *turgida*, Sm. Ditches, etc., frequent.
—— *Zebra*, Kütz. Ditto.
—— *sorex*, Kütz. Ditto.
—— *gibba*, Kütz. Ditto.
—— *ventricosa*, Kütz. Ditto.
CYMBELLA *Ehrenbergii*, Kütz. Ditto.
—— *affinis*, Kütz. Ditto.
—— *cuspidata*, Kütz. Ditto.
AMPHORA *ovalis*, Kütz. Ditto.
—— *minutissima*, Parasitic on Nitzschia sigmoidea, etc.,
at Bury.

AMPHORA *affinis*, Kütz.　Ipswich, Walton Ferry, etc.
—— *hyalina*, Kütz.　　　Ditto.
—— *salina*,　　　　　　Ditto.
COCCONEIS *diaphana*, Common on the coast.
—— *Scutellum*, Ehr.　　Ditto.
—— *Pediculus*, Ehr.　Ditches, etc., frequent.
—— *Placentula*, Ehr.　　Ditto.
COSCINODISCUS *eccentricus*, Ehr.　Ipswich, Walton, etc.
—— *radiatus*, Ehr.　　　Ditto.
—— *concinnus*, Ipswich.
EUPODISCUS *crassus*, In the river Orwell, Ipswich.
—— *radiatus*, Bail ?　　Ditto.
—— *sculptus*,　　　　Ditto, rare.
—— *Argus*, Ehr.　　　Ditto.
TRICERATIUM *Favus*, Ehr.　　Ditto.
ACTINOCYCLUS *undulatus*, Kütz.　Ditto, frequent.
CYCLOTELLA *Kützingiana*, Thw.　Fresh water, frequent.
—— *Rotula*, Kütz.　　　Ditto.
CAMPYLODISCUS *costatus*, Sm.　Mermaid's pits, Bury.
—— *Hodgsonii*, Near Ipswich, rare.
—— *cribrosus*, Sm.　Woodbridge.
—— *parvulus*, Sm.　Felixstow and Ipswich.
SURIRELLA *biseriata*, D. B.　Bury, Ipswich, etc.
—— *linearis*, Bury.
—— *striatula*, Turp.　Felixstow, Walton, Ipswich, etc.
—— *Gemma*, Ehr.　Common on the coast.
—— *Brightwellii*, Sm.　Frequent in fresh or brackish water.
—— *ovata*, Kütz.　Common in ditto.
—— *salina*, Sm.　　　Ditto.
—— *minuta*, D.B.　Bury, etc.
TRYBLIONELLA *gracilis*, Bury, Walton, Ipswich, etc.
—— *marginata*, Ipswich, etc.
—— *acuminata*,　　Ditto.
CYMATOPLEURA *Solea*, Sm.　Fresh water, frequent.
—— *apiculata*,　　　　Ditto.
—— *elliptica*, Sm.　　　Ditto.
NITZSCHIA *sigmoidea*, Sm.　Common in fresh water.
—— *Sigma*, Sm.　Frequent on the coast.
—— *spectabilis*, Near Ipswich.
—— *linearis*, Sm.　Common in fresh water.
—— *Amphioxys*, Sm.　　Ditto.
—— *vivax*, In the river Lark, at Bury.
—— *bilobata*, Sm.　Ipswich.
—— *plana*,　　　Ditto, common.
—— *birostrata*, Sm.　Walton Ferry and Felixstow.
—— *Closterium*, Sm.　Felixstow, Ipswich, etc.
—— *acicularis*, Sm.　Bury, etc.
—— *augularis*, Ipswich.
—— *Tænia*, Sm.　Bury, Felixstow, Ipswich, etc.

NITZSCHIA *lanceolata*, Sm. Walton Ferry.
—— *dubia*, Sm. Ipswich, frequent.
AMPHIPRORA *alata*, Kütz. Common on the coast
—— *vitrea*, Walton and Ipswich.
—— *constricta*, Ehr. Ditto.
—— *elegans*, n. sp. Found at Walton etc., sparingly, by the late Dr. Bleakley, of Norwich.
NAVICULA *rhomboides*, Ehr. About Ipswich.
—— *serians*, Kütz. Ditto.
—— *cuspidata*, Kütz. Ditches about Bury, etc.
—— *firma*, Kütz. Ditto, Ipswich, etc.
—— *Smithii*, Ipswich, rare.
—— *elliptica*, Sm. Bury, Ipswich, etc., frequent.
—— *minutula*, Ipswich, Walton, etc.
—— *Jennerii*, Sm. Ditto.
—— *Westii*, Ditto.
—— *convexa*, Ditto.
—— *elegans*, Ditto.
—— *gibberula*, Kütz. Fresh water, Bury, etc.
—— *ambigua*, Ehr. Ditto.
—— *producta*, Ditto.
—— *Amphisbæna*, Bory. Fresh water, common.
—— *tumens*, Ipswich, etc., common.
—— *punctulata*, Ipswich and Walton, common.
—— *didyma*, Kütz. Common on the coast.
—— *binodis*, Ehr. Bury, etc.
—— *clavata*, Greg. Ipswich.
—— *Lyra*, Walton.
—— *pygmæa*, Walton, Ipswich, etc.
PINNULARIA *major*, Sm. Ipswich, Bury, etc., common.
—— *viridis*, Sm. Plentiful in fresh water.
—— *oblonga*, Sm. Ditto.
—— *peregrina*, Ehr. Ipswich, etc., common.
—— *acuta*, Bury, in ditches.
—— *directa*, Lake Lothing and Ipswich.
—— *radiosa*, Sm. Frequent in fresh water.
—— *gracilis*, Ehr. Bury, etc.
—— *viridula*, Sm. Ditto.
—— *Cyprinus*, Ehr. Ipswich, Walton, etc., common.
—— *divergens*, Bury, etc.
—— *stauroneiformis*, Sm. Bury, Ipswich, etc.
—— *Johnsonii*, Ipswich.
—— *mesolepta*, Ehr. Fresh water, common.
—— *interrupta*, Sm. With the preceding.
STAURONEIS *Phænicenteron*, Ehr. Common in fresh water.
—— *salina*, Ipswich, common.
—— *crucicula*, Ditto, frequent.
—— *anceps*, Ehr. Bury, etc.
—— *linearis*, Ehr. Ditto.

STAURONEIS *pulchella*, Ipswich.
PLEUROSIGMA *formosum*, Sm. Walton, Ipswich, etc.
—— *decorum*, Walton, Felixstow and Ipswich.
—— *speciosum*, Sm. Walton.
—— *rigidum*, Ipswich.
—— *elongatum*, Ditto, frequent.
—— *strigosum*, Sm. Ditto.
—— *quadratum*, Walton, Ipswich, etc., common.
—— *angulatum*, Sm. Ditto.
—— *Æstuarii*, Sm. Ditto.
—— *Balticum*, Sm. Ditto.
—— *Fasciola*, Sm. Ditto.
—— *intermedium*, Ditto, frequent.
—— *acuminatum*, Sm. Felixstow.
—— *prælongum*, Sm. Lake Lothing.
—— *tenuissimum*, Walton.
—— *littorale*, Ipswich and Orford.
—— *Hippocampus*, Sm. Common on the coast.
—— *attenuatum*, Sm. Common in fresh water.
—— *lacustre*, Sm. Bury, Ipswich, etc.
—— *Spencerii*, Sm. Bury and Boyton.
SYNEDRA *pulchella*, Kütz. Bury, etc.
—— *radians*, Sm. Abundant in fresh water.
—— *Ulna*, Ehr. Common in ditto.
—— *affinis*, Kütz. Walton.
—— *fulgens*, Sm. Ipswich.
—— *superba*, Kütz. Ditto.
COCCONEMA *lanceolatum*, Ehr. Fresh water, common.
—— *cymbiforme*, Ehr. Fresh water, frequent.
—— *Cistula*, Ehr. Ditto.
DORYPHORA *Amphiceros*, Kütz. Ipswich, Walton, etc.,
    common.
GOMPHONEMA *constrictum*, Ehr. Fresh water, frequent.
—— *acuminatum*, Ehr. Ditto.
—— *cristatum*, Ralf. Bury, etc., rare.
—— *olivaceum*, Ehr. Common in fresh water.
—— *curvatum*, Kutz. Ditto.
—— *marinum*, Sm. Ipswich, etc.
PODOSPHENIA *Ehrenbergii*, Kütz. Ipswich, etc.
RHIPIDOPHORA *elongata*, Kütz. Ditto.
—— *paradoxa*, Kütz. Ditto.
MERIDION *circulare*, Ag. Frequent in fresh water.
BACILLARIA *paradoxa*, Gmel. Walton, Ipswich, etc., common.
ODONTIDIUM *mutabile*, Fresh water, frequent.
—— *parasiticum*, Walton.
FRAGILARIA *capucina*, Desm. Fresh water, common.
ACHNANTHES *longipes*, Ag. Ipswich, etc., common.
—— *brevipes*, Ag. Ditto.
ACHNANTHIDIUM *lineare*, Sm. Walton.

RHABDONEMA *arcuatum*, Kütz. Ipswich.
—— *minutum*, Kütz. Felixstow, in a Hermit Crab's
  stomach.
DIATOMA *vulgare*, Bory. Frequent in fresh water. Bury, etc.
—— *grande*, Bury.
—— *elongatum*, Bury.
GRAMMATOPHORA *marina*, Kütz. Ipswich.
TABELLARIA *flocculosa*, Kütz. Bury, etc.
—— *fenestrata*, Kütz. Ditto.
AMPHITETRAS *antediluviana*, Ehr. Ipswich.
BIDDULPHIA *aurita*, Breb. Ipswich.
—— *pulchella*, Gray. Walton, Ipswich, etc., common.
—— *Rhombus*, Sm. Ipswich.
PODOSIRA *hormoides*, Kütz. Ipswich.
—— *maculata*, Sm. Ditto, etc., abundant.
MELOSIRA *nummuloides*, Kütz. Ipswich.
—— *Borreri*, Grev. Ditto, common.
—— *varians*, Ag. Frequent in fresh water.
ORTHOSIRA *marina*, Sm. Ipswich, etc., common.
MASTOGLOIA *Smithii*, Thw. Ipswich.
ENCYONEMA *prostratum*, Ralfs. Bury.
—— *cæspitosum*, Kütz. Felixstow.
COLLETONEMA *eximium*, Thw. Walton and Ipswich.
SCHIZONEMA *cruciger*, Lowestoft and Ipswich.
—— *Grevillii*, Ag. Walton, Ipswich, etc., common.

ORD. XCVII. FUNGI.—THE MUSHROOM OR FUNGUS FAMILY.

AGARICUS *phalloides*, Fr. Woods, frequent.
—— *vaginatus*, Bull. Woods and shady places.
—— *muscarius*, L. Fir woods, common.
—— *Mariæ*, Kl. Woodyard at Bury.
—— *rubescens*, Pers. Fir-woods, frequent.
—— *asper*, Pers. Wood at Hardwick.
—— *procerus*, Scop. Pastures, etc., common.
—— *Badhami*, East Bergholt. A.N.H.
—— *cepæstipes*, Sow. Frequent on bark, etc., in stoves.
—— *Clypeolarius*, Bull. Wood at Rougham.
—— *cristatus* Bolt. Common on lawns, pastures, etc.
—— *granulosus*, Batsch. Woods, hedgebanks, etc., common.
—— *melleus*, Vahl. Common on stumps, etc.
—— *hypothejus*, Fr. Fir-wood at Fornham.
—— *rutilans*, Schæff. Stumps in woods, not uncommon.
—— *multiformis*, Schæff. Very common in woods.
—— *acerbus*, Suffolk. A.N.H.
—— *argyraceus*, Bull. Borders of woods, frequent.
—— *personatus*, Fr. Pastures, common.

AGARICUS *nudus*, Bull. Fields, woods, etc., common.
—— *alutaceus*, Pers. Woods about Bury.
—— *nitidus*, Pers. Woods at Great Glemham.
—— *emeticus*, Schæff. Woods, plentiful.
—— *ruber*, Lam. Under oaks in a pasture at Rougham.
—— *fœtens*, Pers. Wood at Great Glemham.
—— *virescens*, Pers. Ditto.
—— *adustus*, a. & b. Pers. Frequent in woods and shady places.
—— *torminosus*, Schæff. Woods. Westley and Great Glemham.
—— *uvidus*, Fr. Woods at Otley.
—— *deliciosus*, L. Not unfrequent in fir-woods.
—— *blennius*, Fr. Beech-woods about Bury.
—— *Volemum*, Fr. Woods. Great Glemham.
—— *quietus*, Fr. Woods, common.
—— *subdulcis*, Bull. Woods at Great Glemham.
—— *turpis*, Suffolk. A.N.H.
—— *glycyosmus*, Fr. Rare. Hardwick heath.
—— *flexuosus*, Pers. Bushy places, Great Glemham.
—— *piperatus*, Scop. Rare. In woods at Otley.
—— *vellereus*, Fr. Woods, frequent.
—— *exsuccus*, Otto. Woods. Great Glemham, etc.
—— *flaccidus*, Sow. Frequent in plantations of fir.
—— *infundibuliformis*, Bull. a. & b. Woods, etc., frequent.
—— *giganteus*, Sow. Among decayed Brakes at Santon Downham.
—— *phyllophilus*, Pers. Woods about Hardwick heath.
—— *canaliculatus*, Schum. Beech wood at Westley.
—— *odorus*, Bull. Sparingly at Santon Downham and Benhall.
—— *candicans*, Pers. Beech wood at Bury.
—— *dealbatus*, Sow. Pastures. Great Glemham, etc.
—— *cerussatus*, Fr. Hardwick heath. Great Glemham.
—— *fimbriatus*, b. Bolt. Hardwick heath.
—— *pratensis*, Pers. Pastures about Bury, Great Glemham, etc.
—— *virgineus*, Wulf. Abundant in pastures.
—— *psittacinus*, Schæff. Pastures, frequent.
—— *separius*, Suffolk. A.N.H.
—— *conicus*, Schæff. Pastures, common.
—— *puniceus*, Fr. Heathy ground at Rougham.
—— *coccineus*, Wulf. Plentiful in pastures.
—— *laccatus*, Scop. a. & b. Common in woods, etc.
—— *sulphureus*, Bull. Dark woods about Bury.
—— *radicatus*, Relh. Frequent at the roots of trees.
—— *pudens*, Great Glemham, rare.
—— *velutipes*, Curt. Common on stumps, etc.

AGARICUS *fusipes*, Bull. Frequent on stumps, etc.
—— *maculatus*, A. & S. Barton park.
—— *confluens*, Pers. Woods, not unfrequent.
—— *dryophilus*, Bull. On oak leaves.
—— *peronatus*, Bolt. Frequent in woods of oak, etc.
—— *oreades*, Bolt. Common in pastures.
—— *conigenus*, Pers. Fir-cones at Hardwick.
—— *Clavus*, Bull. Rotten wood at Bury.
—— *ramealis*, Bull. Decaying sticks at Bury.
—— *Rotula*, Scop. Frequent on dead leaves and sticks.
—— *androsaceus*, L. Dead leaves at Hardwick.
—— *epiphyllus*, Pers. Frequent on leaves in woods, etc.
—— *alcalinus*, Fr. Hedges and woods, frequent.
—— *galericulatus*, Scop. Trunks of trees, frequent.
—— *polygrammus*, Bull. Rotting wood at Bury.
—— *galopus*, Pers. Woods, frequent.
—— *purus*, Pers. Woods, pastures, etc, common.
—— *lacteus*, Pers. Among moss at Westley.
—— *epipterygius*, Scop. On sticks at Hardwick.
—— *corticola*, Bull. On bark at Bury.
—— *Fibula*, Bull. Among moss at Bury.
—— *pyxidatus*, Bull. Great Glemham, rare.
—— *fragrans*, Sow. Sparingly in woods about Bury.
—— *cyathiformis*, Bull. Pastures and woods, frequent.
—— *ostreatus*, Jacq. Trunks of trees. Bury and Great Glemham.
—— *ulmarius*, Bull. On elms at Bury.
—— *palmatus*, Bull. Squared timber, etc., occasionally.
—— *stypticus*, Bull. Common on stumps.
—— *prunulus*, Scop. Pastures, not very common.
—— *leoninus*, Schæff. On sawdust at Bury.
—— *phlebophorus*, Ditm. On sawdust at Stowmarket.
—— *columbarius*, Bull. Pastures at Rougham, etc.
—— *pascuus*, Pers. Hardwick heath.
—— *Sowerbei*, Berk. Pasture at Nowton.
—— *violaceus*, L. Woods at Great Glemham and Westley.
—— *anomalus*, Fr. Wood at Otley.
—— *callochrous*, Pers. Otley, rare.
—— *aureus*, Bull. Frequent on timber.
—— *squarrosus*, Müll. Ditto.
—— *collinitus*, Sow. Woods. Great Glemham.
—— *fastibilis*, Pers. Woods, etc., common.
—— *rimosus*, Bull. Ditto.
—— *geophyllus*, Bull. Ditto
—— *furfuraceus*, Pers. Sticks at Hardwick.
—— *tener*, Schæff. Hedgebanks, pastures, etc., frequent.
—— *involutus*, Batsch. Woods, frequent.
—— *panuoides*, Fr. Rotten timber at Bury.
—— *mollis*, Schæff. Common on timber, etc.

AGARICUS *variabilis*, Pers. On sticks, etc., in woods, common.
—— *speciosus*, Fr. On sawdust at Bury.
—— *Georgii*, With. Woods and pastures, plentiful.
—— *campestris*, L. Pastures, etc., common.
—— *precox*, Pers. Great Glemham.
—— *semiglobatus*, Batsch. Common in pastures, etc.
—— *æruginosus*, Curt. Frequent in various situations.
—— *lachrymabundus*, Bull. Not unfrequent about stumps.
—— *lateritius*, Schæff. Ditto.
—— *fasicularis*, Huds. Common about stumps.
—— *callosus*, Fr. Hardwick heath.
—— *fœnisecii*, Pers. Common at Great Glemham.
—— *atomatus*, Fr. Occasionally about Bury.
—— *gracilis*, Pers. Hedges about Bury.
—— *semiovatus*, Sow. On dung in pastures, frequent.
—— *fimiputris*, Bull. Ditto.
—— *papilionaceus*, Bull. Pastures, frequent.
—— *striatus*, Bull. Near Mermaid's pits, Bury.
—— *titubans*, Bull. Ditto.
—— *disseminatus*, Pers. Very common about stumps, etc.
—— *comatus*, Mull. Roadsides & waste ground, frequent.
—— *picaceus*, Bull. On sawdust at Bury, rare.
—— *atramentarius*, Bull. In a variety of situations.
—— *micaceus*, Bull. Common about stumps, etc.
—— *cinereus*, Bull. Common on sawdust.
—— *niveus*, P. On horse dung, in pastures, etc., frequent.
—— *plicatilis*, Sow. Grassy places, common.
—— *stercorarius*, Bull. Dunghills, frequent.
—— *ephemeris*, Bull. Ditto, common.
—— *radiatus*, Bolt. Frequent on horse dung.
—— *glutinosus*, Schæff. Fir-woods. Great Glemham.
—— *rutilus*, Schæff. Common under Scotch firs.
CANTHARELLUS *aurantiacus*, Wulf. Under fir trees at Rougham.
—— *cibarius*, Fr. Hitcham wood. Tuddenham heath. Great Glemham.
—— *lœvis*, Fr. On Hypnum purum at Hardwick and Bungay.
MERULIUS *corium*, Gr. Common on sticks, timber, etc.
—— *tremellosus*, On timber at Bury.
—— *lachrymans*, Wulf. Common in various situations.
—— *pulverulentus*, Sow. Damp walls, frequent.
SCHIZOPHYLLUM *commune*, Fr. Felled trees at Bury, rare.
DÆDALEA *biennis*, Bull. Stumps, &c. Ickworth park. Great Glemham.
—— *quercina*, L On oak timber, etc., frequent.
—— *betulina*, L. Common on felled trees.
—— *unicolor*, Bull. Frequent on wood.

POLYPORUS *lentus*, Berk. On furze. Rougham and Great
Glemham.
—— *squamosus*, Huds. Common on various trees.
—— *perennis*, L. Fir woods, near West Stow, rare.
—— *varius*, Pers. Frequent on timber.
—— *sulphureus*, Bull. Frequent on trees, especially oak.
—— *hispidus*, Bull. Frequent on various trees.
—— *cæsius*, Schrad. On fir trees near Newmarket.
—— *adustus*, Willd. Not uncommon on timber, etc.
—— *betulinus*, Bull. On a dead birch in Fritton Decoy.
—— *salicinus*, Grev. On willows at Great Glemham, etc.
—— *versicolor*, L. Abundant on felled trees, etc.
—— *abietinus*, Pers. On dead firs, common.
—— *scoticus*, Kl. Common about the roots of trees.
—— *fraxineus*, Bull. On an ash tree at Barton.
—— *dryadeus*, Pers. On ash trees at Hardwick, Barton,
Otley, Great Glemham and Helmingham.
—— *fomentarius*, L. Barton, Benhall, etc.
—— *igniarius*, L. Frequent on various trees.
—— *Ribis*, Schum. On the roots of currant, etc. Great
Glemham.
—— *ferruginosus*, Schrad. Rotting wood. Great Glemham and Bury.
—— *medulla-panis*, Jacq. Great Glemham and Bury, rare.
—— *rapnarius*, On rotting wood at Bury.
BOLETUS *luteus*, L. Fir-plantations, etc., common.
—— *Grevillei*, Kl. Frequent in woods, heaths, etc.
—— *laricinus*, Berk. Under Larch. Nowton and Great
Glemham.
—— *granulatus*, L. Fir-plantations, Bury, Great Glemham, etc.
—— *subtomentosus*, L. Hardwick and Great Glemham.
—— *pachypus*, Fr. Great Glemham.
—— *luridus*, Schæff. Woods at Hardwick.
—— *edulis*, Bull. Great Glemham, Tuddenham, etc.
—— *scaber*, Bull. Woods. Great Glemham, Otley, etc.
—— *cyanescens*, Bull. Wooded ground at Hawstead.
FISTULINA *hepatica*, With. On oak, etc. Bury, Nowton,
and Great Glemham.
HYDNUM *imbricatum*, L. Fir-woods about Bungay.
—— *repandum*, L. Mildenhall. Shrubland park.
—— *compactum*, Pers. About Bungay.
—— *auriscalpium*, L. On Scotch-fir cones, common.
PHLEBIA *mesenterica*, Dicks. Common on wood.
THELEPHORA *palmata*, Scop. Hitcham wood.
—— *laciniata*, Pers. Frequent in fir woods, etc.
—— *rubiginosa*, Schrad. On posts, etc., near the ground,
frequent.
—— *tabacina*, Sow. On hazel, etc. About Bury, rare.

THELEPHORA *hirsuta*, Willd. Very common on wood.
—— *purpurea*, Pers. Frequent on fallen trees, etc.
—— *sanguinolenta*, A. & S. Fir-wood. Bury and Great
    Glemham.
—— *quercina*, Pers. Fallen branches, common.
—— *rufa*, Pers.   Ditto at Hardwick.
—— *cærulea*, Schrad. Common on decaying wood, etc.
—— *gigantea*, Pers. On felled fir-trees at Bury.
—— *Sambuci*, Pers. Very common on elder, etc
—— *ochracea*, Fr. On fallen branches, at Bury.
—— *viscosa*, Pers.   Ditto.
—— *comedens*, Nees. Common on dead branches.
—— *cinerea*, Pers.   Ditto.
—— *acerina*, Pers. Common on maple bark.
CLAVARIA *coralloides*, L. Great Glemham, rare.
—— *stricta*, Pers. On sawdust at Bury.
—— *abietina*, Pers. Fir woods at Westley and Rougham.
—— *muscoides*, L. Hardwick heath.
—— *pratensis*, Pers. Common in pastures, etc.
—— *corniculata*, Schæff. Hardwick heath.
—— *cristata*, Holm. Frequent in woods and shady places.
—— *rugosa*, Bull. Grassy places and woods, common.
—— *inæqualis*, Müll. Meadows about Bury.
—— *helvola*, Pers. Thurston heath.
—— *vermicularis*, Sw. Frequent on lawns and pastures.
—— *amethystina*, Bull. Shrubland park.
COLOCERA *cornea*, Batsch. On squared timber, etc. common.
GEOGLOSSUM *glabrum*, Pers. Rougham and Farnham.
—— *hirsutum*, Pers. About Bury. Rougham heath.
—— *difforme*, Fr. Rougham.
—— *glutinosum*, Pers. Pastures about Bury, frequent.
SPATHULARIA *flavida*, Pers. Fir wood at Bury. Corton heath.
TYPHULA *gyrans*, Batsch. On decaying alder leaves at Bury.
MORCHELLA *esculenta*, L. Woods, hedgebanks, etc. frequent.
—— *semilibera*, D C. Santon Downham, Great Glemham
    and Benhall.
HELVELLA *crispa*, Scop. Woods, frequent.
—— *lacunosa*, Afz. Woods about Bury, frequent.
—— *elastica*, Bull. At Bury and Otley, rare.
VERPA *digitaliformis*, Pers. With Morchella semilibera at
    Great Glemham and Sweffling.
LEOTIA *lubrica*, Scop. Woods at Hardwick & Great Glemham.
PEZIZA *acetabulum*, L. Woods, frequent. Bury, Great
    Glemham, Santon Downham, etc.
—— *reticulata*, Gr. Great Glemham and Hawstead.
—— *badia*, Pers. Hardwick and Great Glemham, rare.
—— *aurantia*, Pers. Horringer, near felled oaks.
—— *cochleata*, Bull. Beech woods at Bury.
—— *cerea*, Sow. On a hot-bed at Bury.

PEZIZA *vesiculosa*, Bull. Dunghills, etc., common.
—— *tuberosa*, Bull. In a damp wood at Hardwick.
—— *cupularis*, L. Beech wood at Bury.
—— *granulata*, Bull. Common on cow dung, etc.
—— *rutilans*, Fr. On the ground, moss, etc., frequent.
—— *humosa*, Fr. Amongst Polytrichum piliferum at
    Fornham.
—— *coccinea*, Jacq. Rotten sticks in hedges, etc., common.
—— *radiculata*, Sow. Wood at Westley. Hedge at Rougham.
—— *hemispherica*, Wigg. Great Glemham.
—— *scutellata*, L. Great Glemham, Santon Downham, etc.
—— *stercorea*, Pers. Very common on cow dung.
—— *albo-spadicea*, Gr. Otley, rare.
—— *virginea*, Batsch. Rotten wood, etc., common.
—— *nivea*, Hedw. In a wood yard at Bury.
—— *calycina*, Schum. Var. c. Common on larch boughs.
—— *cerinea*, Pers. Frequent on wood.
—— *clandestina*, Bull. On wood at Bury.
—— *Schumacheri*, Fr. Ditto.
—— *sulphurea*, Pers. Common on nettle stems, etc.
—— *hyalina*, Pers. On wood at Bury.
—— *villosa*, Pers. Frequent on the dead stems of herba-
    ceous plants.
—— *anomala*, Pers. On wood about Bury, common.
—— *firma*, Pers. In a wood at Fornham.
—— *fructigena*, Bull. On beech-mast, acorns, etc., common.
—— *serotina*, Pers. On stalks of plants. Horringer, Great
    Glemham, etc.
—— *inflexa*, Bolt. Bury and Great Glembam.
—— *Campanula*, Nees. On dead herbaceous stems at Bury.
—— *cyathoeidea*, Bull. Ditto, frequent.
—— *æruginosa*, Pers. Hitcham wood.
—— *pallescens*, Pers. On wood, frequent.
—— *herbarum*, Pers. Common on nettle stems, etc.
—— *cinerea*, Batsch. Common on rotting wood, etc.
—— *compressa*, A. & S. On hard wood, frequent about
    Bury.
—— *sepulta*, East Bergholt. A.N.H.
—— *acicularis*, Bull. About Bury, occasionally.
ASCOBOLUS *furfuraceus*, Pers. On cow-dung. Frequent
    about Bury.
BULGARIA *inquinans*, Pers. On felled trees, etc., common.
—— *sarcoides*, Jacq. Ditto, frequent.
BADHAMIA *nitens*, East Bergholt. A.N.H.
—— *pallida*, Ditto.
—— *fulvella*, Ditto.
DITIOLA *radicata*. A. & S. Ditto.
CENANGIUM *quercinum*, Pers. Common on dead oak branches.

STICTIS *radiata*, L. On wood about Bury.
CRYPTOMYCES *versicolor*, Fr. Common on hard wood.
TREMELLA *foliacea*, Pers. On a felled tree in Ickworth park.
—— *mesenterica*, Retz. Common on furze, wood, etc.
—— *cerebrina*, Bull. On wood. Bury and Great Glemham.
—— *albida*, Sm. Common on wood, etc.
—— *sarcoides*, With. Frequent on wood.
EXIDIA *Auricula-Judæ*, L. On wood at Bury, rare.
—— *recisa*, Ditm. Great Glemham.
—— *glandulosa*, Bull. Frequent on fallen branches.
DACRYMYCES *stillatus*, Nees. Common on wood.
SCLEROTIUM *complanatum*, Tode. Frequent on fallen leaves.
—— *Semen*, Tode. On cabbages, etc., frequent.
—— *fungorum*, Pers. On the gills of blackened Agarics.
    Var. b. Attached to Peziza tuberosa at Hardwick.
—— *durum*, Pers. Common on dead potato stalks, etc.
—— *Pustula*, D.C. Oak leaves at Bury.
—— *salicinum*, D.C. On willow leaves, frequent.
SPERMOEDIA *Clavus*, D.C. Common on wild grasses & rye.
PHALLUS *impudicus*, L. Woods, etc. Santon Downham and
    Great Glemham.
—— *caninus*, Huds. Coddenham, very scarce.
—— *iosmos*, Berk. Sand hills, Lowestoft.
TUBER *cibarium*, Sib. Beech woods at Bury, etc.
—— *moschatum*, Bull. Near a plane-tree in the grounds
    of Lady Cullum, Hardwick, one specimen in 1858.
NIDULARIA *campanulata*, With. Common in various situa-
    tions.
PILOBOLUS *crystallinus*, Tode. On horse-dung at Bury.
—— *roridus*, Bolt. On horse-dung at Great Glemham.
SPHÆROBOLUS *stellatus*, Tode. On rotten wood at Bury.
SPHÆRIA *alutacea*, Pers. Under a fir tree at Westley.
—— *digitata*, L. On decaying wood at Bury.
—— *polymorpha*, Pers. Common on stumps, etc.
—— *Hypoxylon*, L. Abundant on stumps, etc.
—— *carpophila*, Pers. On beech-mast, common.
—— *concentrica*, Bolt. On wood, frequent.
—— *fragiformis*, Pers. On beech bark, common.
—— *fusca*, Pers. On dead hawthorn at Bury.
—— *multiformis*, Fr. Common on dead branches, etc.
—— *serpens*, Pers. Frequent on rotten wood.
—— *deusta*, Hoffm. Rotten stumps, Fornham and Great
    Glemham.
—— *stigma*, Hoffm. Abundant on sticks.
—— *disciformis*, Hoffm. On beech bark, frequent.
—— *aspera*, Fr. On oak branches, common.
—— *verruccæformis*, Ehrh. On hazel at Nowton.
—— *flavo-virens*, Hoffm. Common on wood.
—— *lata*, Pers. Frequent on hard wood.

SPHÆRIA *leucostoma,* Pers. On branches of Pruni, frequent.
—— *cinnabarina,* Tod. Dead trunks, common.
—— *coccinea,* Pers. Branches, common.
—— *Dothidea,* Moug. Frequent on living rose stems.
—— *filicina,* Fr. On brake stems, common.
—— *Junci,* Fr. On Junci, common.
—— *graminis,* Pers. On living grass, common.
—— *Racodium,* Pers. On beech wood, frequent.
—— *pilosa,* Pers. Common on wood.
—— *Peziza,* Tode. On rotten wood at Bury.
—— *sanguinea,* With. On wood, sticks, etc., common.
—— *episphæria,* Tode. Common on Sph. stigma.
—— *mammœformis,* Pers. Wood and sticks, frequent.
—— *stercoraria,* Sow. On dung, near Yarmouth.
—— *spermoides,* Hoffm. Frequent on rotten wood.
—— *pulvis pyrius,* Pers. Common on wood and sticks.
—— *myriocarpa,* Fr. Frequent on wood, etc.
—— *eutypa,* Fr. Frequent on dead branches.
—— *corticis,* Sow. On ash twigs, frequent.
—— *Taxi,* Sow. On yew leaves and twigs, frequent.
—— *Lirella,* M. & N. On dry stems of Spiræa Ulmaria.
—— *acuta,* Hoffm. Plentiful on nettle stems, etc.
—— *complanata,* Tode. Frequent on stems of herbaceous
    plants.
—— *Doliolum,* Pers. Ditto.
—— *culmifraga,* Fr. On culms of various grasses.
—— *herbarum,* Pers. Common on herbaceous plants.
—— *maculœformis,* Pers. Frequent on fallen leaves.
—— *punctiformis,* Pers. Ditto.
—— *herderœcola,* Fr. Common on green ivy leaves.
EUSTEGIA Ilicis, Fr. Common on holly leaves.
LOPHIUM *mytilinum,* Pers. On fir wood at Bury, rare.
CYTISPORA *leucosperma,* Pers. On branches of trees, common.
—— *fugax,* Bull. On willow branches, common.
CEUTHOSPORA *phacidioides,* Grev. Frequent on holly leaves.
—— *Lauri,* Sow. On laurel leaves, at Ickworth.
PHOMA *Pustula,* Pers. Common on oak leaves.
DOTHIDEA *ribesia,* Pers. On dead branches of the red currant.
—— *typhina,* Pers. Common on the culms of grasses.
—— *rubra,* Pers. On sloe leaves, etc., common.
—— *betulina,* Fr. On birch leaves, frequent.
—— *Ulmi,* Duv. Extremely common on elm leaves.
—— *Robertiani,* Fr. On leaves of Geran. Robertianum.
RHYTISMA *salicinum,* Pers. On willow leaves.
—— *corrugatum,* Ach. On old barn doors.
—— *Acerinum,* Pers. On sycamore and maple leaves.
PHACIDIUM *coronatum,* Fr. On leaves of oak, etc., frequent.
—— *dentatum,* Schm. On fallen oak leaves.
HYSTERIUM *pulicare,* Pers. Frequent on trunks of trees.

HYSTERIUM *Fraxini*, Pers. Common on ask twigs.
—— *rugosum*, Fr. Common on smooth bark.
—— *pinastri*, Schrad. Common on pine leaves.
—— *culmigenum*, Fr. On grass, frequent.
—— *Rubi*, Pers. On dead bramble stems, Bury.
BATARREA *Phalloides*, Woodw. Sand hills about Bungay
and elsewhere in the county, very rare.
GEASTER *coliformis*, Dicks. Sandy banks at Mettingham.
Lane near Bexley Common.
—— *fornicatus*, Huds. Occasionally met with.
—— *striatus*, D.C. Sandy ground about Yarmouth and
Bungay.
—— *Bryantii*, Berk. Sandy bank at Bury. About Bungay.
—— *limbatus*, Fr. In a fir wood at Bury.
—— *mammosus*, Chev. Sandy ground, Santon Downham.
—— *rufescens*, Pers. Fir woods at Bury, plentiful.
—— *hygrometricus*, Pers. In a sandy wood at Bury.
BOVISTA *nigrescens*, Pers. Heaths at Rougham and Santon
Downham.
—— *plumbea*, Pers. Common in dry pastures, etc.
LYCOPERDON *giganteum*, Batsch. Frequent about Bury,
Great Glemham, etc., etc.
—— *cœlatum*, Bull. Frequent in pastures about Bury.
—— *gemmatum*, Batsch. Abundant in pastures, etc.
—— *pyriforme*, Schæff. On stumps, not common.
—— *atropurpureum*, Vitt. Hardwick, in a wood.
TULOSTOMA *mammosum*, Fr. Old walls at Bury, frequent.
SCLERODERMA *vulgare*, Fr. Great Glemham.
—— *verrucosum*, Bull. Common in hedgebanks and woods.
LYCOGALA *Epidendrum*, L. Frequent on rotten wood.
RETICULARIA *umbrina*, Fr. Frequent on wood.
ÆTHALIUM *septicum*, L. Common on sawdust, tan, etc.
SPUMARIA *alba*, Bull. Frequent on grass. Ickworth park.
Nowton, Great Glemham, etc.
DIDERMA *vernicosum*, Pers. On grass, occasionally.
—— *spumaroides*, Fr. On moss at Hardwick.
—— *globosum*, Pers. On oak leaves at Bury.
DIDYMIUM *hemisphœricum*, Bull. On leaves at Bury.
—— *squamulosum*, A. & S. On rotten twigs at Bury.
—— *nigripes*, Lk. On rotten wood at Bury.
PHYSARUM *nutans*, Pers. Ditto, frequent.
—— *hyalinum*, Pers. Ditto.
—— *sinuosum*, Bull. On tan in a hot-house at Lady Cullum's, Hardwick.
—— *album*, Nees. On decaying wood, frequent.
CRATERIUM *minutum*, Leers. On leaves, Hardwick heath.
—— *leucocephalum*, Hoffm. On leaves at Bury.
STEMONITIS *fusca*, Roth. Common on rotten wood.
—— *ovata*, Pers. On decaying wood, frequent at Bury.

STEMONITIS *papillata*, Pers. On decaying wood.
ARCYRIA *punicea*, Pers. Frequent on rotten wood.
—— *incarnata*, Pers. On rotten sticks in Woolpit wood.
—— *nutans*, Bull. On rotten wood at Bury, frequent.
TRICHIA *fallax*, Pers. Rotten wood at Great Glemham.
—— *clavata*, Pers. Ditto, at Bury.
—— *turbinata*, With. Ditto.
PERICHÆNA *populina*, Fr. On fallen poplars at Bury.
TRICHODERMA *viride*, Pers. On fallen trees, etc., common.
RACODIUM *cellare*, Pers. In old cellars, common.
ERYSIPHE *communis*, Schl. Common on various plants.
—— *bicornis*, Lk. On maple and sycamore, common.
—— *guttata*, Schl. On hazel, frequent.
CHÆTOMIUM *elatum*, Kunz. Common on straw.
ILLOSPORIUM *roseum*, Fr. On Parmelia parietina at Hardwick.
ISARIA *farinosa*, Fr. On dead pupæ about Bury, rare.
CERATIUM *hynoides*, A. and S. Frequent on rotten wood.
ASCOPHORA *Mucedo*, Tode. On bread, etc., common.
HYDROPHORA *stercorea*, T. On the dung of various animals.
MUCOR *ramosus*, Bull. On decaying Fungi, etc., frequent.
—— *Mucedo*, L. Very common on various substances.
—— *caninus*, Pers. On the dung of dogs and cats.
—— *fusiger*, Lk. On the gills of decaying Agarics. Bury.
EUROTIUM *herbarum*, Lk. On various decaying bodies.
HELICOSPORIUM *pulvinatum*, Fr. On wood at Bury.
HELMINTHOSPORIUM *macrocarpum*, Grev. On rotten wood
    at Bury.
—— *velutinum*, Lk. On dead stalks of Symrnium at Bury.
—— *nanum*, Nees. On wood at Bury.
CLADOSPORIUM *herbarum*, Lk. On various decaying sub-
    stances, abundant.
ASPERGILLUS *candidus*, Lk. Ditto.
—— *glaucus*, Lk. On bread, cheese, etc., common.
—— *roseus*, Lk. On damp paper at Bury.
STACHYLIDIUM *diffusum*, Fr. Frequent on rotting plants.
BOTRYTIS *parasitica*, Pers. On various Cruciferæ.
—— *effusa*, Gr. On spinach leaves at Barton.
—— *infesta*, The potato blight, too common.
PENICILLIUM *crustaceum*, Fr. Very common on decaying
    substances.
—— *candidum*, Lk. Frequent on ditto.
DACTYLIUM *dendroides*, Fr. On decaying Fungi at Bury.
SPOROTRICHUM *sulphureum*, Grev. On corks, hampers, etc.,
    in cellars.
—— *laxum*, Lk. Frequent on various bodies.
OIDIUM *monilioides*, Lk. On leaves of grasses at Bury.
—— *erysiphoides*, Fr. On leaves of various plants.
—— *leucoconium*, Desm. On rose leaves at Bury.
SPORENDONEMA *Casei*, Desm. On cheese, frequent at Bury.

SPORENDONEMA *muscæ*, Fr. Common on dead flies.
SEPEDONIUM *chrysospermum*, Lk. Common on decaying
Fungi.
FUSISPORIUM *gris um*, Fr. Common on fallen leaves.
EPOCHNIUM *fungorum*, Fr. Common on Thelephoræ.
PSILONIA *rosea* Berk. On fir cones at Hardwick.
TUBERCULARIA *vulgaris*, Tode. Very common on dead
sticks, etc.
FUSARIUM *tremelloides*, Grev. On decaying nettle stems,
common.
NÆMASPORA *crocea*, Pers. Dead beech trees at Bury.
SEPTORIA *Ulmi*, Kunz. Common on elm leaves.
—— *Badhami*, East Bergholt. A.N.H.
AREGMA *bulbosum*, Fr. On bramble leaves, common.
—— *mucronatum*, Fr. On rose leaves at Great Glemham.
TORULA *antennata*, Pers. On sticks at Bury.
—— *ovalispora*, Berk. On wood at Bury.
—— *herbarum*, Lk. Frequent on the stems of Umbelliferæ.
PODISOMA *Juniperi communis*, Fr. On living Juniper at
Bury.
PUCCINIA *Graminis*, Pers. On leaves and culms of corn
and grasses.
—— *striola*, Lk. On Carices, Junci, etc.
—— *Polygonorum*, Lk. On Polygona, frequent.
—— *Menthæ*, Pers. Common on Mints.
—— *Umbelliferarum*, D.C. On Umbelliferæ, frequent.
—— *Saxifragarum*, Schl. On Adoxa moschatellina at
Hardwick.
—— *Prunorum*, Lk. On plum leaves, common.
—— *Fabæ*, Lk. On the horse-bean, frequent.
—— *Saniculæ*, Grev. On Sanicula europæa at Hardwick.
—— *Buxi*, D.C. On box-leaves at Hardwick.
ÆCIDIUM *Ari*, Berk. On Arum maculatum at Bungay.
—— *rubellum*, Pers. On dock leaves, frequent.
—— *Menthæ*, D.C. On mints, frequent.
—— *Compositarum*, Mart. On various Compositæ.
—— *Ranunculacearum*, D.C. Common on Ranunculaceæ.
—— *Berberidis*, Pers. Common on Berberis vulgaris.
—— *albescens*, Gr. On Adoxa moschatellina at Hardwick.
—— *Grossulariæ*, D.C. On gooseberry leaves, common.
—— *laceratum*, Sow. On hawthorn, rare.
—— *Euphorbiæ*, Pers. Frequent on spurges.
—— *Urticæ*, D.C. Common on nettles.
UREDO *segetum*, Pers. Common on various cereals.
—— *Caries*, D. C. Common on wheat.
—— *linearis*, Pers. Common on corn and grasses.
—— *Rubigo*, D.C. Ditto.
—— *olivacea*, D.C. On Carex riparia at Oulton.
—— *Labiatarum*, D.C. On the Labiatæ, frequent.

132

UREDO *Vincæ*, D.C. On Vinca major at Bury and Bungay.
—— *compransor*, Schl. On various Compositæ.
—— *suaveolens*, Pers. On Cnicus arvensis, common.
—— *Umbellatarum*, Johnst. On various Umbelliferæ.
—— *Ranunculacearum*, D.C. Frequent on Ranunculaceæ.
—— *Rosæ*, D.C. On rose leaves, common.
—— *effusa*, Str. On leaves, petioles, etc., of roses.
—— *Ruborum*, D.C. On brambles, frequent.
—— *Potentillarum*, D.C. On Rosaceæ, frequent.
—— *Leguminosarum*, Lk. On Leguminosæ, frequent.
—— *Quercus*, Brond. On oak leaves, Bungay.
—— *candida*, Pers. On Cruciferæ, very common.
—— *Euphorbiæ*, Reb. On spurges, frequent.
—— *cylindrica*, Str. On Populi, frequent.
—— *Saliceti*, Sch. On Salices, frequent.
—— *Caprearum*, D.C. On sallows, common.

GENERAL INDEX.

134

Bittersweet 59.
Blackberry 25.
Black-bryony 86.
Blackthorn 25.
Bladder-wort 55.
Blechnum 105.
Blinks 30.
Blue-bottle 48.
Blysmus 91.
Bog-bean 57.
Bog-orchis 83.
Bog-rush 90.
Boletus 124.
Bonnemaisonia 114.
Boraginaceæ 57.
Borago 59.
Borkhausia 50.
Borrera 113.
Botrychium 105.
Botrytis 130.
Bovista 129.
Brachypodium 100.
Bramble 25.
Brassica 7.
Briza 101.
Brome-grass 100.
Broom 20.
Broom-rape 60.
Bromus 100.
Brooklime 62.
Brookweed 54.
Bryonia 30.
Bryopsis 114.
Bryum 107.
Buck-bean 57.
Buckthorn 19.
Buckwheat 72.
Bugle 68.
Bugloss 58.
Bulbochæte 115.
Bulgaria 126.
Bull-rush 79.
Bunium 37.
Bupleurum 35.
Burdock 47.
Bur-marigold 47.
Burnet 26.
Burnet-saxifrage 25.
Bur-reed 79.
Butcher's-broom 87.
Butomus 81.
Butter-bur 42.

Buttercups 3.
Butterwort 55.
Cakile 10.
Calamagrostis 97.
Calamintha 65.
Calicium 109.
Calithamnion 115.
Callitriche 75.
Callitrichineæ 75.
Calluna 53.
Calocera 125.
Calothrix 115.
Caltha 3.
Calycifloræ 19.
Calystegia 57.
Camelina 8.
Campanula 52.
Campanulaceæ 52.
Campion 12.
Campylodiscus 117.
Cantharellus 123.
Caprifoliaceæ 39.
Capsella 9.
Caraway 34.
Cardamine 6.
Carduus 47.
Carex 92.
Carlina 48.
Carpinus 76.
Carrot 37.
Carum 34.
Caryophyllaceæ 11.
Catabrosa 103.
Cat-mint 66.
Caucalis 37.
Celandine 2, 5.
Celastraceæ 19.
Celery 34.
Cenangium 126.
Centaurea 48.
Centaury 48, 56.
Centranthus 41.
Centunculus 54.
Ceramium 114.
Cerastium 13.
Ceratium 130
Ceratophyllum 74.
Cetraria 113.
Ceuthospora 128.
Chærophyllum 37.
Chætospora 114.
Chætophora 116.

Chætomium 130.
Chaffweed 54.
Chamomile 44.
Chara 113.
Charlock 8.
Cheiranthus 5.
Chelidonium 5.
Chenopodiaceæ 69.
Chenopodium 70.
Cherry 25.
Chervil 37.
Chickweed 14.
Chicory 51.
Chlora 56.
Chondrus 114.
Chorda 114.
Chordaria 114.
Chroolepus 116.
Chrysanthemum 43.
Chrysosplenium 32.
Chylocladia 114.
Cichorium 51.
Cicuta 34.
Cineraria 46.
Cinquefoil 26.
Circæa 29.
Cistaceæ 10.
Cladium 90.
Cladonia 113.
Cladosporium 130.
Cladostephus 114.
Clary 64.
Clavaria 125.
Clematis 1.
Clinopodium 65.
Closterium 116.
Clover 21.
Club-moss 104.
Cnicus 47.
Cocconeis 117.
Cocconema 119.
Cochlearia 8.
Cock's-foot 101.
Colchicum 89.
Collema 112.
Colletonema 120.
Colt's-foot 42.
Columbine 3.
Comarum 26.
Comfrey 59.
Compositæ 42.
Conferva 115.

THE END.